GeoJournal Library

Volume 129

Series Editor

Barney Warf, University of Kansas, Lawrence, KS, USA

More information about this series at https://link.springer.com/bookseries/6007

Weifeng Li · Lingqian Hu · Jason Cao
Editors

Human-Centered Urban Planning and Design in China: Volume I

Urban and Rural Planning

 Springer

Editors
Weifeng Li
University of Hong Kong
Hong Kong SAR, China

Lingqian Hu
School of Architecture and Urban Planning
University of Wisconsin–Milwaukee
Milwaukee, WI, USA

Jason Cao
Humphrey School of Public Affairs
University of Minnesota, Twin Cities
Minneapolis, MN, USA

ISSN 0924-5499   ISSN 2215-0072   (electronic)
GeoJournal Library
ISBN 978-3-030-83858-4   ISBN 978-3-030-83856-0   (eBook)
https://doi.org/10.1007/978-3-030-83856-0

This Springer imprint is published by the registered company Springer Nature Switzerland AG
The registered company address is: Gewerbestrasse 11, 6330 Cham, Switzerland

Preface

This book explores a more human-centered development pathway associated with the ideological shift from "quantity" to "quality" growth in the new era of Chinese urbanization. Sustainable urban and rural planning should be "people-centered" and concerned about urban-rural coordination. The authors argue that successful urban and rural development in China should promote social equity, culture diversity, economic prosperity, and sustainable built form.

This book prompts Chinese urbanists to reconsider and explore a sustainable and people-first planning approach with Chinese characteristics. The breadth and depth of this book is of particular interest to the faculty members, students, practitioners and the general public who are interested in subjects like urban and regional planning, rural planning, housing and community development, infrastructure planning, climate change and ecological planning, environmental planning, social equity, and beyond.

This book dealing with human-centered urban planning and development, rural planning and urban-rural coordination in China is part of a two-volume set. Volume II discusses human-centered urban design and placemaking, human activities, and urban mobility.

This volume includes 17 peer-reviewed and rigorously edited papers, which were presented at the 11th and 12th International Association for China Planning (IACP) Conferences held, respectively, June 16–18, 2017 at Harbin Institute of Technology in Harbin, China and June 30-July 1, 2018 at Xi'an University of Architecture and Technology in Xi'an, China, on the theme Human-Centered Urban Planning and Design in China.

Part I: Urban Planning and Development

In Chap. 1, Hui Li et al. examined the way to the leading and regulating role of its spatial resources configuration in Urban Growth Boundary (UGB) to ensure national and regional ecological security and harmonious socioeconomic development in chapter "On Chinese Approach to Theoretical Study of Urban Growth Boundary

(UGB)". They discussed the Chinese approach to UGB theory from the perspective of multidisciplinary integration using research accomplishments, representative figures, regions and case studies, comparative analysis using the qualitative classification method, layer-by-layer detailed method, growth method, exclusive method, comprehensive method, and the like principles and models. They put forward the application of the integrated comprehensive method to the approach to define the future urban growth boundary. The research findings provided scientific basis for thorough research and real UGB theory demarcation.

Le Che et al. interpreted the wisdom of the city's core idea from the inspects of information technology, knowledge economy, and ecological sustainable development, and conducted a relevance study to verify the interactive circular relationship between knowledge, wisdom, and ecology in Chap. 2. The DICEF model was proposed to guide the space development of smart city from spatial distribution, spatial competition, spatial interaction, spatial evolution, spatial flow, and to facilitate space demand and spatial planning strategy of development trend.

In Chap. 3 Yan Huang claimed that, throughout China, traditional ecological wisdom had played an important role of guiding human behavior, and had proven safe and effective over the years, in achieving ecological and aesthetic excellence. People from ancient China took the idea of ecology to a philosophical and aesthetical level. In accordance with landscape practices, the author introduced the application of tradition Chinese ecological wisdom in a current sustainability environment by explaining issues such as traditional sustainable ecological principles, surrounding environment and residents instead of squandering natural resources, complicated high technology. More specifically, the study explored how landscapes could maximize surrounding resources and adopt the simplest and most effective method as well as how to influence people's ecological attitudes and daily behavior through environmental design.

Chun Li et al. combined the analysis of remote sensing image interpretation to characterize the changing patterns of green space in Wuhan, China from 1995 to 2015 in Chap. 4. Results revealed that green space control planning accounted for the progress of green space change. The urban growth boundary (UGB) failed to achieve its management and control targets. While forests had recovered efficiently in some areas, urban expansion in the city's suburbs had invaded and destroyed a significant amount of farmland and grassland. The authors further discussed the underlying causes of unproductive control planning by considering the stage of development in Midwest China.

In Chap. 5, Mu-En Change et al. examined the relationship between vegetation cover, land use and urban hear island (UHI) effects. They calculated the Fractional Vegetation Cover (FVC) and value of land surface heat of Taipei City within Taiwan using the satellite image data of Landsat 8 in 2015. They found that the most serious urban hear island effect was the commercial and industrial land use in special use district with FVC less than 30% and no UHI was found for national park and scenic area with FVC both high than 35%. This research could serve as the reference for the urban design and the spread control of UHI in the future.

Huifen Huang and Dazhi Gu summarized the principle of evolution of urban spatial morphology and the influence of natural ecological elements toward urban spatial form in Hefei City through induction–deduction method and literature review in Chap. 6. They further analyzed the deficiencies based on the present situation of urban spatial morphology and discussed the optimizing strategy and future developing trend from the ecological perspective, thereby providing guiding suggestions on planning urban spatial form to other cities.

In Chap. 7, Linglan Bi et al. examined the ecological adaptive mechanism of traditional human settlements in Sichuan Tibetan areas based on cultural perspective. This study focused on the socio-cultural development, experiences from the ecological adaptation allied to sustainable development in Sichuan Tibetan areas. Based on the interactive relationship of the evolutionary process of the human settlement environment, the corresponding mechanism was verified, and the limitations of the application were clarified. The ecological adaptation mechanism and its critical point were analyzed by combining different misconceptions of human living environment constructions in different development stages and spatial scales.

Chong Liu and Ying Han revealed the influences of religion and Buddhism temples in specific on the urbanization over the Mongolia region in Qing Dynasty in Chap. 8. They found the urban areas surrounding the temples, as temples were usually the initial focal point of urban development, gradually expanded and Mongolia's nomadic residential community fundamentally changed. They also analyzed the reasons behind the creation and growth of these urban towns based on the examples of Dorenol, Xilinhot, Guihua City, Kulun Town, and Uliastai Town.

In Chap. 9 Shiwei Shao et al. proposed a simple and effective method of understanding the characteristics and distribution of urban functional areas, based on POI feature vector and network kernel density. Based on the term frequency-inverse document frequency (TF-IDF) algorithm, the POI eigenvectors of the basic units were calculated. The K-means clustering method was applied to cluster all basic units into K functional zones based on POI eigenvectors and various functional zones were initially identified. Finally, the network kernel density estimation and Kriging interpolation method were applied. Using Wuhan as an example, they discovered the distribution of functional zones in the city and the results showed this method being effective to define the functional urban zones instantaneously and clearly.

Pei Chen and Yaping Huang evaluated the degree of industry-city integration in Hubei Province according to the degree and status of industry-city integration, including the composite dimension, hardware foundation, software environment, and interactive dimension in Chap. 10. Given the national-wide transformation of High-Tech Industrial Development Zones, the analytical approach applied in Hubei case provided a good reference to other mid-west regions of China.

Part II: Rural Planning and Urban-Rural Coordination

In Chap. 11, Wenting Jiang et al. investigated the transformation of urban-rural relationship and the role played by small and medium town development in the urban-rural relationship transformation in Shaoxing City. The case study suggested that the key to urban-rural coordination is to promote the life quality of rural residents and to narrow the gap between urban and rural areas. The study might provide reference for urban-rural coordination planning in similar districts where small and medium towns gather.

Haiyan Pang and Yonghua Li analyzed urban space policy in the process of rapid urbanization using Geospatial and Remote Sensing approaches in Chap. 12. The space policy and urban spatial growth mode correlation analysis suggested that: (1) city structure optimization under the action of adjustment policy, (2) city function upgrade under the leading policy, and (3) balance urban and external space under the action of reverse policy. They concluded that space policies influenced spatial growth by forming the shape, changing the timeliness, and shifting the stage.

In Chap. 13 Yuan Huang and Yanxiao Pan studied China's policies for responding to the effects from climate change caused by rapid urbanization. China released a wide range of urban and rural planning policies addressing aspects of urban community operation and layout optimization, ecological afforestation, infrastructure protection, and energy-efficient design, buildings, and transport. The authors detailed adaptation policies that emerged recently and analyzed existing problems of urban and rural planning in adaptation policies, as well as proposed modes of alternative development.

Fan Yang et al. discussed the contemporary issue of "Rural Revitalization" originated from the policies of "New Rural Construction" and "Beautiful Village", in Chap. 14. In their study, the relationship between urban planning and rural planning was seen as imperative for the academic planning circles to refresh these policies of urbanization. It revealed that rural communities are self-organized, law-enforced and based on the structure of collective units. Rural land rights is a fundamental framework for deciding the production capacity and lifestyle of villagers, at the same time a key factor for individual decision-making by villagers, and a critical issue for understanding China's dialectical urbanization and rural development cycle. They discussed the possible methods of responding to the above-mentioned aspects and attempted to argue that rural planning education would be one of the major components of urban planning education.

In Chap. 15 Shuting Yan et al. assessed 16 national characteristic towns, in Hubei province which was representative of the central and western regions in China. Factors such as GDP growth, population loss, and lack of innovation industries were analyzed and identified as the main obstacles in the development of characteristic towns. The authors also compared Japanese prefectures' urbanization processes and development indicators to that of Chinese provinces. The output suggested, through gain and losses analysis in the first and second stages of rapid growth of urbanization in the northeast region of Japan, that improving traffic network, attracting labor talent, and ensuring the proper construction of industrial structure might help solve the problems posed by the characteristic town development in Hubei region.

Yu Guo et al. stated that the unit project cost of rural cultural heritage elements is an important driving force to preserve rural cultural landscape, in Chap. 16. The traditional Shiyang Town Linpan was chosen as the research object in the gravity irrigation agriculture cultural heritage region of Dujiangyan important irrigation province. Lowest construction cost was estimated through the optimal solution model. The authors concluded that the shaping of Linpan landscape heritage should fully reflect the characteristics of rural culture and the analysis of the cost of landscape construction based on the optimal solution model was more suitable for shaping the rural cultural heritage in western Sichuan.

Finally, in Chap. 17, Jiehao Zhu et al. looked into Urban-rural integration and coordinated development as a strategy for rural planning in contemporary China. The plan for rural revitalization, declared during the 19th National Congress of the CPC as a key for rural development of China, tasked scholars in urban and rural studies to increase their focus on the development of rural regions. The authors addressed deficiencies in current Chinese rural planning in three fields, namely, the related legislation, systems of governance, and technical means. This study identified that coherent and reasonable rural planning methodologies can only be developed through a joint consideration of the legislation, governance mechanisms, and planning technologies.

As the proceedings of the 11th and 12th IACP annual conferences, this book aims to develop a sustainable and people-first planning approach with Chinese characteristics. The discussion will contribute to the advancement of urban planning and design in China as well as the world.

Hong Kong SAR, China   Weifeng Li, Ph.D.
Milwaukee, WI, USA   Lingqian Hu, Ph.D.
Minneapolis, MN, USA   Jason Cao, Ph.D.

Contents

Part I
Urban Planning and Development

Chapter 1
On Chinese Approach to Theoretical Study of Urban Growth Boundary (UGB)

Li Hui, Hua Yiling, Li Yadong, Zhang Yu, Li Zhiying, Fan Honghong, and Ren Jiahao

1.1 Research Background

Globally, natural disasters have been growing in intensity and frequency, and were identified as the development theme of the 2016 World Urban Development Report by the United Nations Human Settlements Program (UN-Habitat) and the United Nations Environment Program (UNEP) to build "compatible, secure, resilient and sustainable" cities of the United Nations Human Settlements Program (UN-Habitat) and the United Nations Environment Program (UNEP). In the face of urban growth and land use changes, it is even more important for us to protect agricultural land, biodiversity, quality underground water supplies, and fragile forests and coastal areas. In terms of urban space, it is directly reflected in the change of landscape spatial pattern, i.e., the changing application functions of all kinds of land, and in terms of implementation, it is dependent on the demarcation of various area spatial boundaries.

Both urban and rural artificial construction areas, or natural and semi-natural urban ecological system composed of rivers, coastal line, farm land, massif and country park should have clear boundary, to control and guide urban spatial configuration. Correspondingly, the boundary includes natural, semi-natural and urban town and village boundary, or it is known as the boundary between construction area and non-construction area (Andres & Elizabeth, 1995). How to plan all kinds of spatial configurations in a unified manner, and scientifically delimit corresponding

L. Hui (✉)
College of Forestry and Landscape Architecture, South China Agricultural University, Guangzhou, China
e-mail: ydlihui@qq.com

H. Yiling
Department of Architecture, Waseda University, Tokyo, Japan

L. Yadong · Z. Yu · L. Zhiying · F. Honghong · R. Jiahao
School of Architecture and Urban Planning, Yunnan University, Yunnan, China

© Springer Nature Switzerland AG 2021
W. Li et al. (eds.), *Human-Centered Urban Planning and Design in China: Volume I*,
GeoJournal Library 129, https://doi.org/10.1007/978-3-030-83856-0_1

boundary for implementation, is an important guarantee for the role of urban ecological system functions & benefits, so as to cope with dramatic urbanization of large and medium sized cities during rapid urbanization. Urbanization invades the surrounding ecological land and leads to the waste of land resources, as has become the key and difficult points for us to make sustainable development strategies.

The first Urban Growth Boundary (UGB) emerged in Salem city, USA in 1976, as "the delimit boundary between urban land and rural land" to resolve the UGB resolved the urban planning and management conflict between Salem City and the two counties of Marion and Polkby determining that the land within the boundary could be used for urban land development, while the land outside could not (Wu & Lu, 2014). After that, western scholars supplemented the definition of Urban Growth Boundary. Richard (1991) believes that to decide on the urban growth boundary is to draw a line outside the city to curb unchecked expansion of the urban area" (Andres & Elizabeth, 1995). Bengston David et al. (2004) define the Urban Growth Boundary as an important boundary to tell an urbanized area from its surrounding suburb ecological reserve space, as is marked by the government on the map, whose implementation is guaranteed by zoning and other policy tools (Gerrit & Lewis, 2001). The most famous urban growth boundary is located in Portland metropolitan area in Oregon, USA. By 1999 than 100 cities/rural areas in America have established an Urban Growth Boundary, and three states (Oregon, Tennessee and Washington) have Urban Growth Boundaries that covered their entire states.

1.2 Domestic and International Research Progresses of Urban Growth Boundary

In western countries, scholars have explained the implications of urban growth boundaries from different perspectives. From the perspective of demarcating its spatial scope, a metropolitan area should have an intuitive, clear-cut boundary, such as a river, coastline, farmland, massif and suburban park which the city should not go across or encroach upon as it expands (Andres & Elizabeth, 1995). From the perspective of boundary composition, UGB is a useful regional planning tool to control and guide both rural and urban boundary composition. The urban boundary has been defined by Howard from the perspective of a city as an independent, continuous boundary that encircles the urban area to limit growth. In contrast, the suburban boundary has been defined by Benton and Paul from the perspective of nature, as a defined boundary to inhibit suburban land from encroaching on nature. The two boundaries can be overlapped or separated (Andres & Elizabeth, 1995). From the perspective of time–space evolution, the two boundaries have both dynamism and permanency (Gerrit & Lewis, 2001).

In China, scholars have paid extensive attention to urban growth boundaries since 2007, but they have very different understanding. Some follow the landscape ecology thinking and take the urban growth boundary as a regional boundary eliminating natural space (including farmland, forest land, and water area) or suburban zones (Wu & Lu, 2014). Others, from the perspective of urban development demand, believe the urban growth boundary includes preserved space to meet future demand as the city expands (Huang & Casella, 2007). Another view believes the urban growth boundary includes a "rigid" boundary to ensure ecological security by protecting non-construction land and elastic which can be adjusted along with growing urbanization (Minghua & Xiaoqing, 2008).

Some define the boundary for broadly controlling urban development as a generalized urban growth boundary, while a boundary for planned urban construction land is considered an urban growth boundary in its narrow sense (Long et al., 2009). Another perspective is that the urban growth boundary is the limits of built-up area delimited according to the law between the urban service boundary and the green belt. Only the land within the boundary can be used for future urban construction, while the land outside the boundary is reserved for agricultural development or preserved as open space for ecological purpose (Miao, 2010). Similarly, Bin Live and Qingzheng Xu believe that although the urban growth boundary represents the specific line between construction and non-construction spaces , the urban growth boundary in nature reflects the balance of growth and constraint, demand and supply, and impetus and hindrance, which can be taken as isolines of functional urban region, ecological function area and agricultural function area. It is recommended that the timeliness of growth boundary shall be based for division work. Prospective urban growth boundary shall be revised after a number of years, and permanent vigilance urban growth boundary (urban ecological security bottom line not to be developed).

In summary, UGB is the line between the land for urban construction, and the land reserved or protected for other purposes. UGB is a technique, tool, and policy measure to prevent unchecked, disorderly urban growth and provides cities with a clear, approved space for growth during a set time period.

1.2.1 International Research

Relevant international research mainly focuses on how to scientifically and precisely define the urban planning scope, demarcate the development space of metropolitan areas, and set the technique methods for urban growth boundaries (Tables 1.1 and 1.2).

1.2.2 Domestic Research

The concept of urban space growth boundary was introduced in China in 1990. In 2006, the Urban Planning-Making Method put forward the initial urban space growth boundary. In 2014, the National New Urbanization Plan 2014–2020 reiterated the importance of limiting urban development while optimizing the spatial expansion

Table 1.1 International qualitative demarcation method of urban growth boundary

Method designation	Main approaches and procedures	Representative figures or areas	Main principles
Qualitative demarcation method of urban growth boundary	①Identify regional development issues, ②collect and analyze urban land use growth data, ③anticipate the population and land use growth demand in the future (20–30 years), ④demarcate urban growth boundary.	Frey, Mary	It is easy to provide infrastructures and public services facilities, avoid natural reserves while protecting important scenic spots, important agricultural, historical, and cultural areas, and resource-rich zones.
Layer by layer detailed demarcation of urban growth boundary	①Putting forward four possible modes for future urban growth boundary; ②they are known as "growth concept", each of which corresponds to one urban development scenario; ③by comprehensive comparison, we have preliminarily formed the urban growth boundary; ④in combination with the land utilization status quo, service center and main street location, environment sensitive area and location not suitable for development, we have detailed processing of the urban growth boundary, to demarcate the growth boundary form.	Knaap and Arthur(1992)	Comparison principles for all scenarios: ①newly added land shall be made as close as possible along traffic routes and distributed close to business centers to improve land utilization; ②we take the least possible amount of ecological open space including water reserves outside the boundary and forest and farm land; ③urban form should be applicable to comprehensive traffic system including rail transportation, bus, private car, bike and walking; ④urban development shall keep community partitioned and coordinated with surrounding cities and avoid continuous development;we provide diversified habitats for residents in the area.

Table 1.2 International quantitative demarcation of urban growth boundary

Method designation	Methods and principles	Main models and methods	Representative figures or areas
Growth method	This method takes urban construction land as a continuous growing organic body, in combination with urban growth model simulation and references simulation results to demarcate the urban growth boundary, and mainly used to demarcate elastic growth boundary. Common thinking is to extract growth rate, strength, and direction etc. parameters making use of historic construction land growth data, in combination with population and employment to predict the land use scale and simulate the urban land form in the future	①Based on cellular automaton (CA) dynamic simulation, ②SLEUTH model, ③converted land use and effect model (CLUE), ④ Artificial Neural Network Model (ANN),⑤Mixed model including Logistic regression model, Markov chain and CA model.	Peter et al. (2002) (Iran Teheran metropolitan area), Silva and Clarke (2002) (Lisbon and Porto, Portugal), Pijanowski et al. (2009), Amin and Bryan (2011)
Excluding method	We should exclude unsuitable or non-applicable land due to limited construction condition, and sensitive ecological environment ctc. to identify the maximum amount of urban construction land; it is mainly used to demarcate rigid growth boundary.	With 45 m spatial resolution data, we design three scenarios, each of which has different protection of park, wetland, beach, forest, agricultural land and water area; growth management zone scope, whether including new planned road and stops, slope limit strength, and by eliminating unavailable development area we demarcate the urban growth boundary under different ecological security requirements.	Claire et al. (2003) (Washington—Baltimore metropolitan area)

structure and required the demarcation of boundaries for megacities. Chinese scholars concentrate their UGB research on the domestic and international UGB differentiation, interpretation of urban space growth boundary, and the demarcation method of urban space growth boundaries (Table 1.3).

Table 1.3 Domestic quantitative demarcation of urban space growth boundary

Method designation	Methods and principles	Main models and methods	Representative figures or areas
Growth method	This method takes urban construction land as continuous growing organic body, in combination with model simulation for urban growth and utilizes simulation results to demarcate urban growth boundaries. It is mainly used to demarcate elastic growth boundary. Common thinking is to extract growth rate, strength, direction etc. parameters, making use of historic construction and land growth data, in combination with population and employment to predict the land use scale, and simulate the urban land form in the future.	①GIS spatial technology; ②with SPSS to build local dynamic model; ③SLEUTH model;	He et al. (2010) (Beijing), Zhang et al. (2013) (Lake Dianchi Basin on Yunnan-Guizhou Plateau), Li et al. (2014) (Lake Dianchi Basin)
Excluding method	We should exclude unsuitable or non-applicable land due to limited construction condition or ecological environment sensitive etc.to identify the maximum value of urban construction land. It is mainly used to demarcate rigid growth boundary.	Ecological adaptability assessment, ecological sensitivity analysis, landscape ecological security pattern, scenario analysis, and analytic hierarchy process (AHP)	Zhou et al. (2007), Jie et al. (2016), Li et al. (2014)

(continued)

Table 1.3 (continued)

Method designation	Methods and principles	Main models and methods	Representative figures or areas
Comprehensive method	We should consider urban growth restriction factors and use quantitative methods like game mechanism, spatial growth simulation and grid analysis method to integrate the prediction of the growth trend. It is mainly used to demarcate the elastic growth boundary.	①First, from the restricting factors of urban land growth, we determine the natural factors hindering urban growth, and the order of importance for each effect. We select the leading factor to evaluate urban growth control conditions from the ecological perspective; ②with spatial analysis software as technical support, in combination with historical data, transportation, policy etc., we use the factors to simulate the urban growth, and research into the development direction of urban land; ③Based on the results from the quantitative evaluation of construction land suitability and urban scape prediction, functional positioning and spatial structure envision, we make use of optimal selection of high land construction suitability, to determine the urban growth boundary	Cai et al. (2017), Jing et al. (2016) (Anqing), Wang and Gu (2017) (Suzhou), Chengri (2012)

1.2.3 Summary

International and domestic growth boundaries have different backgrounds. Western countries mainly represented by the United States primarily use growth boundaries to handle the expansion of the urban area and the receding central cities caused by suburbanization. Compared with that in the United States, suburban land is even more compact and limited in China. Here the growth boundary is mainly to cope with the drastic expansion of large and medium cities during urbanization and to protect them from encroaching on the surrounding ecological land and wasting land resources. Therefore, the urban growth boundary of the Urban and Rural Planning Law and the Urban Planning-Making Regulations is based on urban land use evaluation and research on urban development direction, to comprehensively determine the urban space growth boundary for land use that urban development may involve.

Urban growth boundary is based on previous research, but it is seldom aimed at functional, urban spatial regions that guarantee national and regional ecological security and coordinated socioeconomic development. Most researchers explore from their own perspective, focusing on a few research regional spaces as a whole, but most research remains conceptual and theoretical, and in actual application there are overlapping spaces, and the boundary is blurred.

1.3 Chinese Approach to Urban Space Growth Boundary

The urban space growth boundary in this research refers to a demarcated functional area and the corresponding boundary for coordinating urban ecological system service function within the scope of the urban and rural spatial area. This ensures maximum ecological benefits and no spatial overlapping, so as to guarantee national and regional ecological security, a coordinated socioeconomic spatial development pattern, and to provide scientific and technological methods for regional planning, policy making, management and implementation. Corresponding to natural ecological system, and artificial and semi-artificial ecological system, it can provide urban ecological system service function that can meet and sustain the condition and process and the like urban ecological system service functions. Specifically, it includes ecological protection, functional role of the urban system, and staple food supply and like the main ecological system service functions, which are corresponding to the "ecological red line" of forestry and environmental protection departments, "urban growth boundary" of housing and urban & rural development, and 'basic farmland protection boundary" of agricultural department; breaking the conventional fragmentation. Under the theory and framework of giving the overall role of urban system services function, explore "integration of multiple regulations" scientific and rational methods.

1.3.1 Demarcating the Rigid Urban Growth Boundary

Considering China's national condition, the "rigid" urban growth boundary mainly refers to the ecological red line and basic farmland protection boundary.

Ecological Red Line Demarcation

In terms of space, we can construct the ecological protection network from the three aspects of functional ecological importance, ecological environmental sensitivity/fragility, and environmental disaster/pollution danger to ascertain ecological corridor, radiation channel, ecological strategic nodes and the ecological security pattern components' spatial distribution. In terms of quantity, we can use ecological carrying capacity research to ascertain the minimum area of the ecological red line within a certain space area. From space and quantity, we can establish the strict management and control of key ecological function areas, biological diversity areas, ecological environmental sensitivity areas, and fragility areas, so as to maintain ecological security, guarantee ecological system function, and support the sustainable development of the economy and the society.

Basic Farmland Protection Boundary Demarcation

We can ascertain the time–space configuration scheme and demarcate the basic farmland protection boundary. With ecological system service value, topographic slope, soil erosion modulus, geological disaster rate and continuous rate, and the like farmland ecological quality representing indexes, we construct the indexes for farmland to be listed as basic farmland, and put forward basic farmland demarcation comprehensive appraisal method; in combination with the above material flow research, synthesizing spatial form and ecological security, ecological carrying capacity and the like factors, and based on GIS platform, we realize all index quantitative analysis and farmland quality comprehensive index estimation.

1.3.2 Demarcating the Elastic Urban Growth Boundary

Comprehensively considering growth pressure, growth overflow possibility, infrastructure and future supply capacity, distance to different service centers, distance to expressway, railway and main highway and factors driving urban growth through natural resources, ecological sensitivity, infrastructures, public services facilities level and so on, we can calculate the bearing capacity of population and socioeconomic activities, i.e. evaluation of overall bearing capacity as important reference factors of growth boundary by comprehensively considering growth pressure, growth overflow possibility and so on. Applying cellular automaton based SLEUTH model, and through dynamic monitoring and mechanism exploration of urban development, we predict the city's future growth trend, form scenarios for urban expansion under different parameters, and evaluate the environmental impact of different scenarios. In combination with the ecological red line results, the basic farm land protection boundary demarcation results, and natural and historic boundaries, we can finally decide on the urban growth boundary.

The three above urban growth boundary demarcations are not separated, but comprehensively consider the balance between growth and constraint, demand and supply, and impetus and resistance of a functional urban region, ecological function area, and agricultural function area. Based on the timeliness of growth boundary, we demarcate "rigid" boundary, i.e. permanent vigilance urban growth boundary (urban ecological security bottom line not to be developed), and "elastic" boundary, i.e. along with the growth of the city, we can adjust the space growth boundary, with preserved space to meet the demand for the future expansion of the city.

1.4   Chinese Practice of Demarcating "Rigid" Urban Space Growth Boundary

The research takes the Nujiang River Basin (middle section) (Fig. 1.1) as a typical case: The place is known for its high mountains and deep valleys, drastic elevation, intense vertical altitude, and complicated climate, soil and vegetation; conditions for various natural and geographic landscapes, ecological environment, and biology. From valley bottom to mountain top, it is home to scarce species, relic species, and endemic species. Due to narrow terrains and steep valley slopes, its ecological environment is very fragile, so urban and rural construction, agricultural production, industrial and mining enterprises and transportation, and water and electricity facilities have been constrained. The factors affecting the ecological environment quality are complex. Therefore, ecological essentiality is evaluated with the single factor superposition method and the logic rule combination method together. Based on the complexity between evaluation factors, we use logic rules to establish ecological essentiality analysis criteria, and based on this, we have judgment and analysis. With geographic information system technology applied, we have the final importance evaluation results.

1.4.1   Procedures for Ecological Essentiality Evaluation

Based on the characteristics of this area, we will have soil erosion sensitivity evaluation and habitat sensitivity evaluation, to conduct analysis evaluation, to synthesize and have the ecological important evaluation results of this area. Its procedures go as follows (Fig. 1.2)

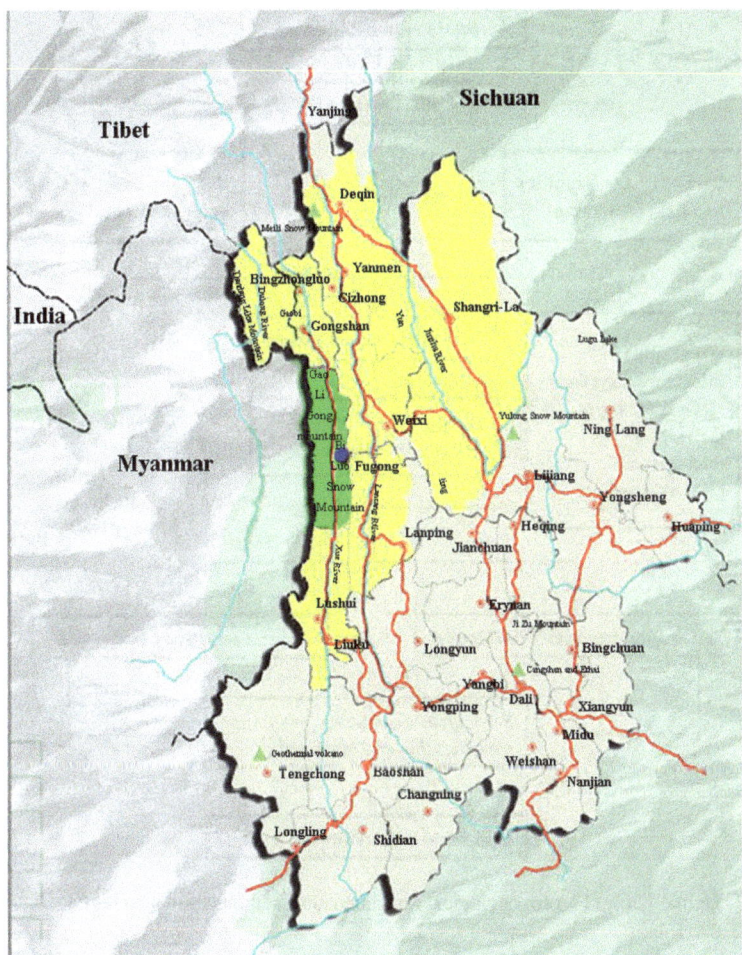

Fig. 1.1 Location of the investigated region in the northwest Yunnan

1.4.2 Soil Erosion Sensitivity Evaluation Method

Single Factor Selection and Evaluation Method

Factors involved in soil erosion include climate, hydrology, landform, soil and vegetation, and conservation of soil and water, and other natural and man-made factors. Referring to the general soil erosion equation, in combination with the characteristics of this research area, and 3S technology feasibility research, we determine the research area oil erosion sensitivity evaluation factors including water erosion (R), landform (LS), soil geology factor (K), and land cover factor (C) (Figs. 1.3, 1.4, 1.5, 1.6).

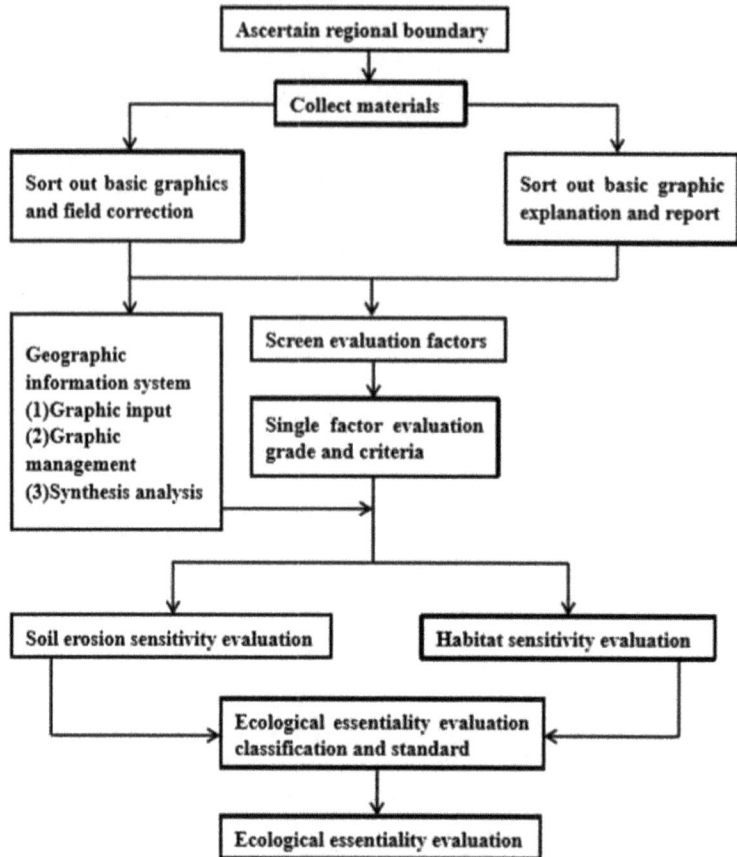

Fig. 1.2 Opinion flow chart for ecological fundamentality in the investigated region

Assessment Criteria

All influencing factors have classification and standard for soil erosion sensitivity evaluation, according to the specifications for soil erosion sensitivity evaluation compiled by the State Administration for Environmental Protection, in combination with the actual situation of the research area, as shown in Table 1.4.

Soil Erosion Sensitivity Comprehensive Evaluation Method

If we analyze a single factor to have soil erosion sensitivity, it will show the function of one factor, but not comprehensively show the variation of the research area. This research makes use of a comprehensive soil erosion sensitivity evaluation index, whose equation is:

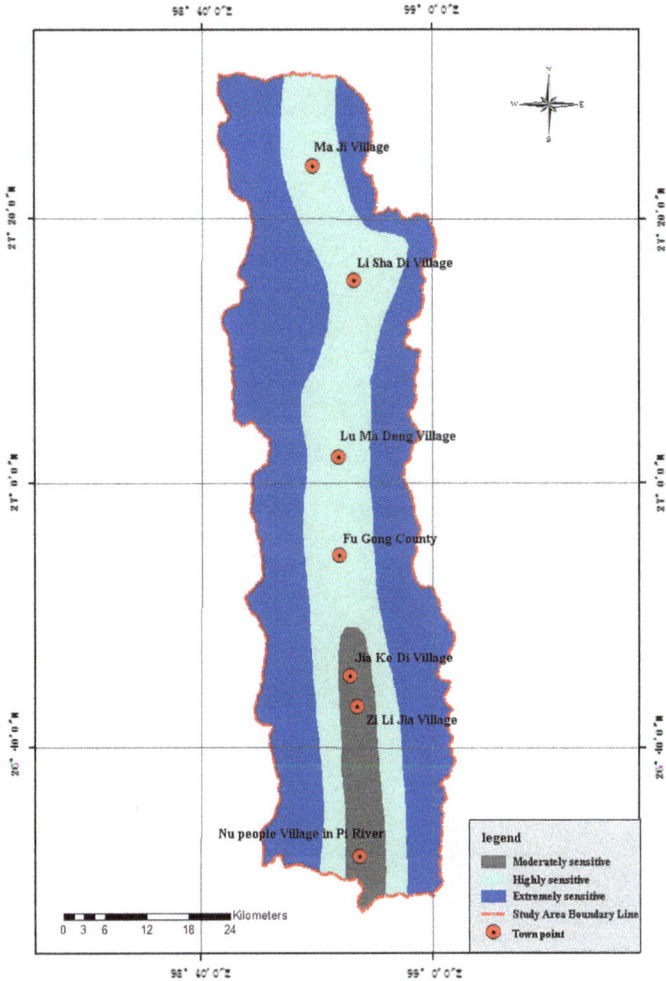

Fig. 1.3 Sensitivity grade subarea for rainfall eroding factor in the investigated region

$$SS_j = \sqrt[4]{\prod_{i-1}^{4} C_i}$$

where SS_j is the soil erosion sensitivity index of space unit j; and C_i is the sensitivity grade of factor i.

To conduct the overlay analysis according to the above equation, we use soil erosion sensitivity of the research area analyzed from single factor, and make use of the GRD module of ARC/NFO. First, we convert each layer of data into grid data. To ensure data precision, reduce data redundancy, and improve stacking velocity, we use rational polygon integration. We have overlapping map layers through the

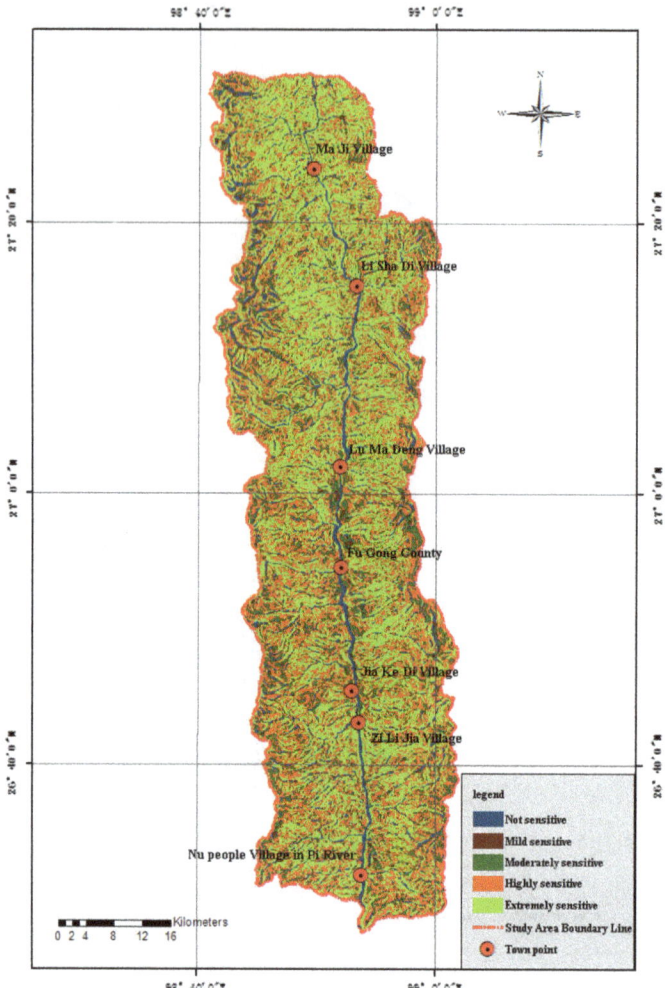

Fig. 1.4 Sensitivity grade subarea for gradient factor in the investigated region

UNION command, and finally overlay the four map layers. According to the classification standard, we assign values to the classification results of the four factors;
and through ARCGIS software we compile the equation against the above equation. After operation we have the results, and use the classification standard (SS)
to conduct evaluation classification, to have evaluation results. To this end, after
classification mapping, we have comprehensive evaluation distribution map for soil
erosion sensitivity (Fig. 1.7).

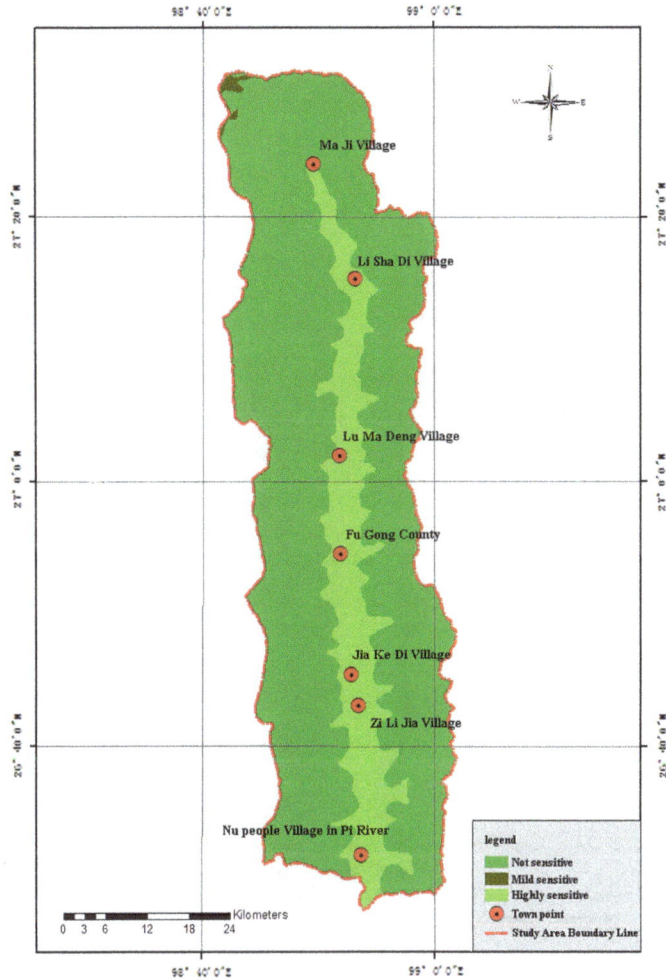

Fig. 1.5 Sensitivity grade subarea for soil factor in the investigated region

Evaluation Results for Soil Erosion Sensitivity in the Investigated Region

GIS overlay analysis and statistics indicate that most of the research area is a medium or highly sensitive area, and there is no extremely sensitive area (Tables 1.5,1.6 1.7). The medium sensitive area is 2059.15 km², accounting for 75.31% of the total area, and it is mainly distributed around the region whose altitude is more than 2,000 m, while the highly sensitive area is 455.22 km², accounting for 16.65% of the total area, and it is mainly distributed in the region whose altitude is less than 2,000 m. The insensitive area is 14.88 km², accounting for 0.54% of the total area, and mainly distributed where altitude is more than 4,000 m. The lightly sensitive area is 205 km²,

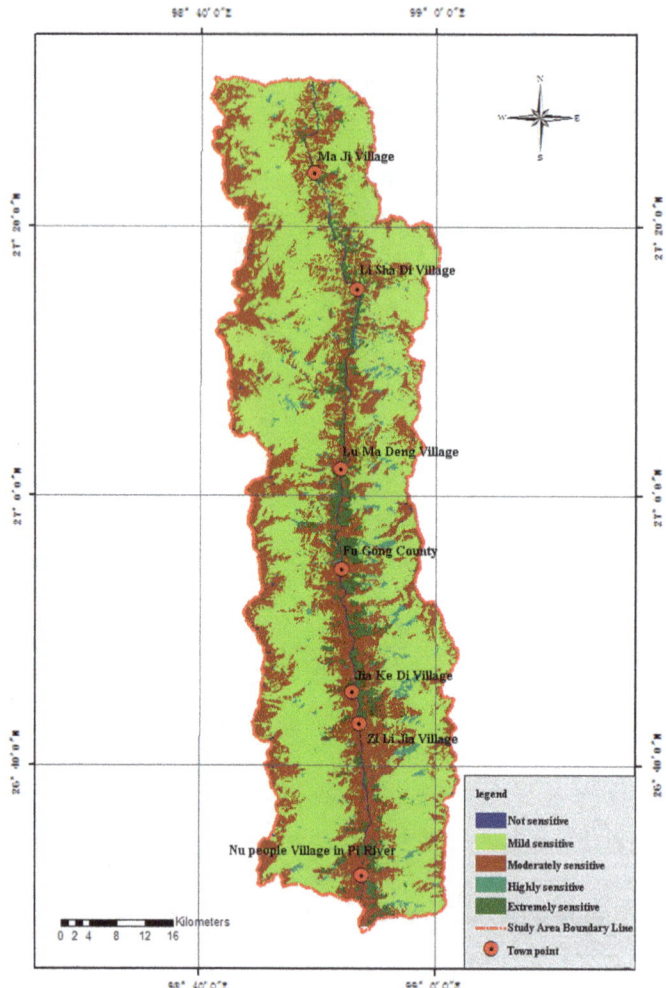

Fig. 1.6 Sensitivity grade subarea for the earth's surface factor in the investigated region

accounting for 7.50% of the total area, and mainly distributed along the branches of the Nujiang River with altitude above 2,000 m.

1.4.3   Habitat Sensitivity Evaluation Method

Biodiversity Protection Sensitivity Evaluation Method

Table 1.4 Classification standard for factor influencing soil erosion sensitivity

Classification	Insensitive	Lightly sensitive	Intermediately sensitive	Highly sensitive	Extremely highly sensitive
Precipitation (mm)	≤900	900~1000	1000~1200	1200~1500	1500~2000
Soil texture (k)	≤0.099	0.099~0.160	0.160~0.228	0.228~0.329	0.329~0.475
Landform (slope)	0~8°	8~15°	15~25°	25~35°	>35°
Land coverage	Glaciers and permanent snow, towns, irrigated farm land, river, lake, pond, naked rock, gravel land, rural residential area	Improved grassland, artificial grassland, open forest land, natural grassland, forest land	Vegetable field, shrub wood, other field, afforest land	Wild grass ground, slash	Dry land, bare land, other unused land, irrigable land, mud flat, special use land, independent industrial land
Classification and assignment (C)	1	3	5	7	9
Classification standard (SS)	1.0~2.0	2.1~4.0	4.1~6.0	6.1~8.0	>8.0

* k is soil erosion factor, ranging 0~1

The habitat sensitivity evaluation of the research area mainly refers to biodiversity protection sensitivity. Biodiversity involves three levels, i.e., genetic diversity, diversity of species, and ecosystem diversity. Among the three biodiversity protection levels, ecosystem diversity is the carrier of genetic diversity and diversity of species. According to the ecological features and biodiversity distribution law of the research area, the ecosystem diversity with vegetation types as main representation is the specific embodiment of habitat sensitivity in the area. Therefore, according to the actual situation in this area, the habitat sensitivity evaluation is mainly based on the species richness of the ecosystem, and the distribution of national and provincial protected species.

Assessment Criteria

According to the specifications for habitat sensitivity evaluation compiled by the State Administration for Environmental Protection, the classification and standard for habitat sensitivity of biology diversity based on species protection is shown in Table 1.6.

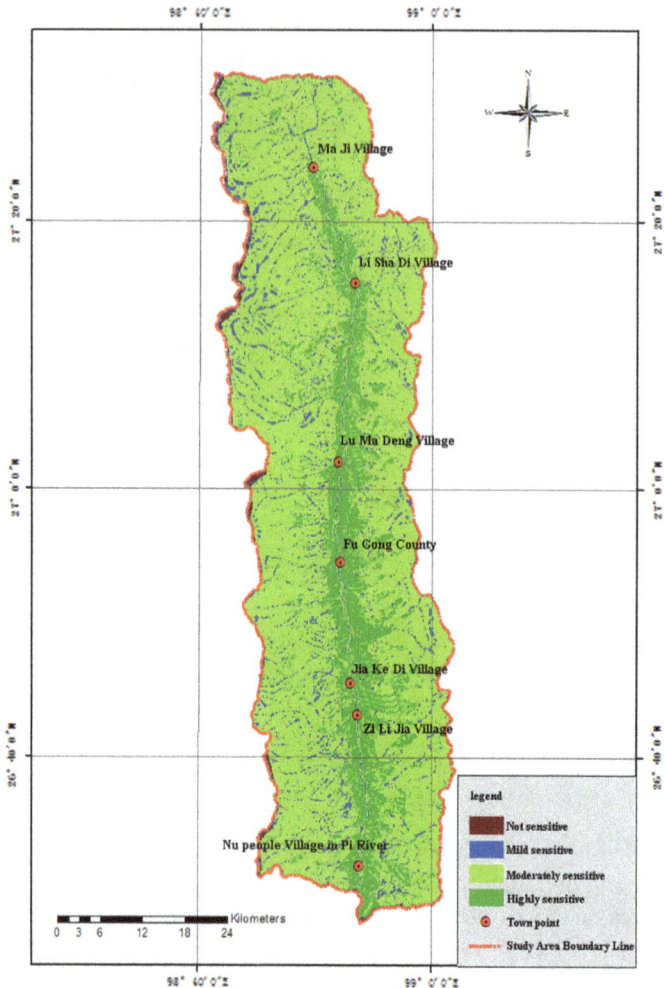

Fig. 1.7 Opinion for soil erosion sensitivity in the investigated region

Meanwhile, according to the richness of the above protected species in one ecosystem, the ecosystem sensitivity of the research area is classified.

Habitat Sensitivity Evaluation Results

Taking different ecosystem type sensitivity grade as the classification standard, we reassign value to the interpreted vegetation types to generate a new map layer, and with GIS spatial analysis and statistics function, we conduct the evaluation.

Table 1.5 Degree and distributing area of soil erosion sensitivity in the investigated region

Sensitivity grade	Number of patches (pcs)	Patches proportion (%)	Patch area average (km²)	Area (km²)	Area proportion (%)
Insensitive	267	3.13	0.06	14.88	0.54
Lightly sensitive	3911	46.40	0.05	205.00	7.50
Intermediately sensitive	1753	20.80	1.17	2059.15	75.31
Highly sensitive	2492	29.57	0.18	455.22	16.65
Total	8428	100%		2756.45	100%

Table 1.6 Classification standard opinion for habitat sensitivity of biology diversity

National and provincial protection species	Habitat sensitivity grade
National level 1	Extremely sensitive
National level 2	Highly sensitive
Other national and provincial protection species	Intermediately sensitive
Other regional protection species	Lightly sensitive
Unprotected species	Insensitive

Data source Ecology and Geobotany Research Institute, Yunnan University, Research Report of Ecological Function Zoning of Yunnan [R], 2004:52.

Table 1.7 Distributing area for habitat sensitivity degree in the investigated region

Sensitivity grade	Number of patches	Patches proportion (%)	Area (km²)	Area proportion (%)
Insensitive	1177	22.45	194.44	7.05
Intermediately sensitive	636	12.13	856.74	31.06
Highly sensitive	2088	39.83	923.14	33.47
Extremely highly sensitive	1341	25.58	783.63	28.41

The results indicate that: The area of habitat insensitive area of the research area is 194.44 km² accounting for 7.05% of the total area, and it is mainly distributed along river valleys where people inhabit; the area of medium sensitive area is 856.74 km², accounting for 31.06% of the total area, and it is mainly distributed around the area whose altitude is more than 3,000 m; the area of highly sensitive area is 923.14 km², accounting for 33.47%, and it is mainly distributed around the area whose altitude is 1,300 ~2,900 m; the extremely sensitive area is 783.63 km², accounting for 28.41% of the total area, mainly distributed around the area whose altitude is less than 2000 m and around 2900 m (see Table 1.6 and Fig. 1.8).

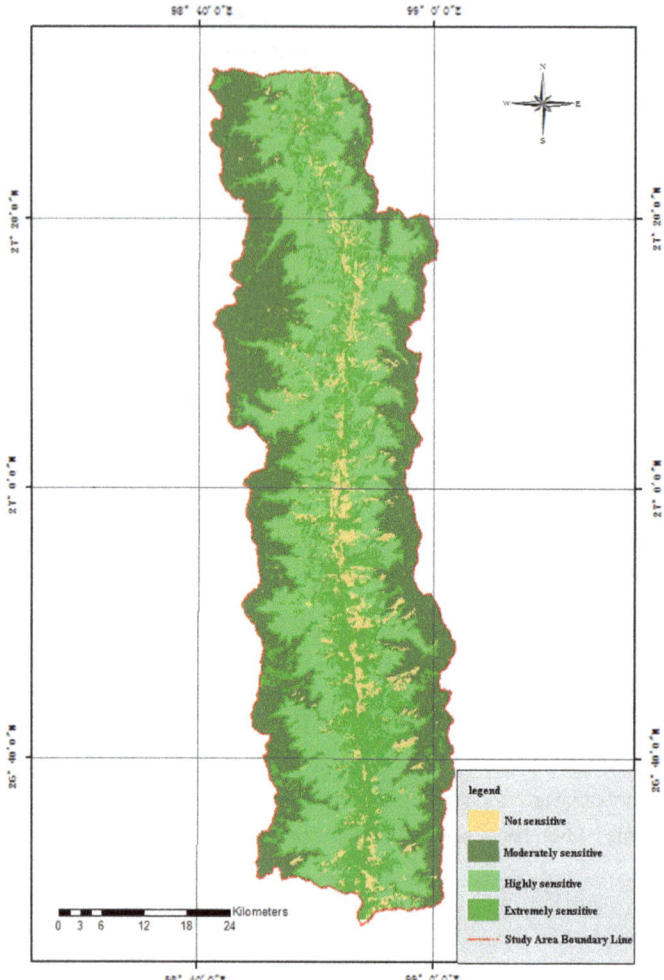

Fig. 1.8 Opinion for habitat sensitivity in the investigated region

1.4.4 Ecological Essentiality Evaluation Method

According to the alpine and gorge region complex ecosystem features of the research area, ecological essentiality mainly includes the combined effect of soil erosion sensitivity and habitat sensitivity.

Table 1.8 Classification standard opinion for ecological essentiality

Sensitivity Importance	Extremely highly sensitive habitat	Highly sensitive habitat	Intermediately sensitive habitat	Lightly sensitive habitat	Insensitive habitat
Soil erosion extremely highly sensitive	Extremely important	Extremely important	Very important	Very important	Important
Soil erosion highly sensitive	Extremely important	Extremely important	Very important	Important	Important
Soil erosion intermediate sensitive	Very important	Very important	Important	Commonly important	Commonly important
Soil erosion lightly sensitive	Very important	Important	Commonly important	Commonly important	Not important
Soil erosion insensitive	Important	Important	Commonly important	Not important	Not important

Assessment Criteria

Based on the combination from habitat extremely sensitive and soil erosion extremely sensitive area to habitat insensitive and soil erosion insensitive area, we can divide the sensitivity importance into five levels: extremely important, very important, important, important and not important (see Table 1.8).

Ecological Essentiality Grade Evaluation Results

According to the above evaluation criteria, we have the grade distribution of the five ecological essentiality levels in the research area (see Table 1.9 and Fig. 1.9).

Table 1.9 Opinion for ecological essentiality grade

Protection grade:	Number of patches (patches)	Patches proportion (%)	Area (km^2)	Area proportion (%)
Not important	24	0.30	1.09	0.04
Commonly important	2073	26.30	144.53	5.29
Important	3078	39.05	961.61	35.22
Very important	1543	19.57	1310.97	48.01
Extremely important	1165	14.78	312.18	11.43

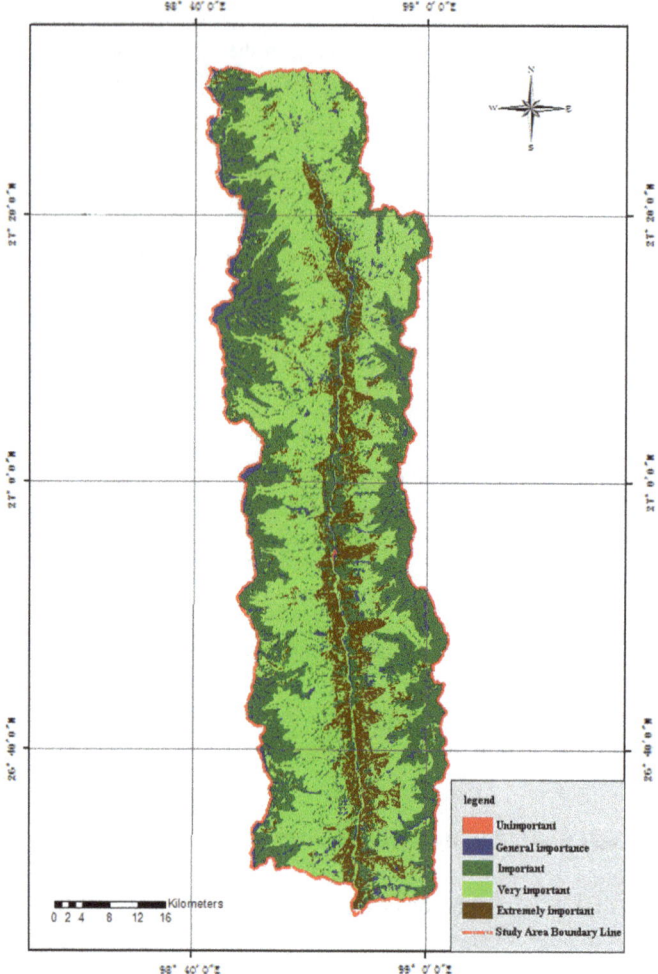

Fig. 1.9 Opinion for ecological essentiality in the investigated region

Evaluation Results

Ecological essentiality mainly considers the comprehensive ecological service functions including water source conservation, water and soil conservation, biodiversity protection, and disaster prevention and reduction. Considering the features of the alpine and gorge region, we analyzed soil erosion sensitivity and habitat sensitivity together, and this is reflected in the comprehensive evaluation results.

Habitat sensitivity is closely related to vegetation and wildlife distribution, and the habitat sensitivity area is always the most ecologically essential area, so the ecological essential area and its essentiality should be consistent with the habitat sensitivity, and the area with the richest biodiversity. Soil erosion sensitivity is mainly related to precipitation, soil texture, and slope. A highly sensitive soil erosion area is always an area of high ecological essentiality. The consequent ecological essentiality evaluation results will provide the scientific basis for demarcating a "rigid" urban growth boundary in the research area.

1.5 Conclusion and Discussion

1.5.1 Conclusion

International experience and domestic practice tell us that the urban space growth boundary includes dual "rigid" and "elastic" functions. The "rigid" growth boundary (vigilance boundary), which is determined from the perspective of protecting non-construction land around a city, is the ecological security bottom line for sustainable urban development It has strong operational significance, and embodies the ultimate scale of the city. Conversely, the "elastic" growth boundary (anticipated boundary) comes from the perspective of growth of construction land, is a dynamic boundary of urban space expansion over different periods, and reflects the stage features of urban development. The principle of urban planning shall shift from expansion planning to limiting urban boundary and optimizing spatial structure, and considering the urban development stage to demarcate the "rigid" and "elastic" urban development boundaries.

Based on the above investigation, this research puts forward an applied, integrated comprehensive method for urban growth boundary demarcation. The integrated comprehensive method refers to comprehensive qualitative and quantitative consideration of integrating the ecological red line, urban growth boundary, and basic farmland protection boundary, with coordination and integration to the scientifically defined urban function boundary. (1) Based on sensitivity structure, fragile process, essential function, and ecological bearing capacity, we can conduct an ecological essentiality evaluation to demarcate the land which is most essential in ensuring ecological system service function, biodiversity and habitat protection. They are considered as the ecological red line, whose area proportion shall be more than 30% of total land area, as is the "rigid" boundary for urban ecological protection. (2) We can use the Land Evaluation and Site Assessment (i.e. LESA) method to define priority farmland from the perspectives of ecosystem service value, farmland natural quality, utilization conditions, space form and ecological security, and consider population flow in this area to research and calculate its ecological bearing capacity, to finally demarcate the basic farmland protection boundary, as the "rigid" boundary for urban material supports . (3) We can refer to the layer-by-layer detailed method of *Portland Plan 2040* to demarcate the urban growth boundary. First, we put forward four possible modes for the urban growth boundary in the future; through comprehensive comparison, we preliminarily form the urban growth boundary, and in combination with the land utilization status quo, service center and main street location, environment sensitive area and location not suitable for development, we have detailed processing of the urban growth boundary. Thereinto, the four possible urban growth modes are called "growth concept", and each is corresponding to one urban development scenario. Based on different scenarios, we determine the ultimate development mode, and demarcate the urban growth boundary, as the "elastic" boundary for urban construction and development.

1.5.2 Discussion

According to the institutional reform of the State Council in March 2018, the planning function of the newly organized Natural Resources Ministry will go to the National Development and Reform Commission, which mobilizes specialists to compile the Main Function Area Plan. This will surely end the case that all departments have fragmented division in spatial planning, to form new land spatial planning considering all functions and integrating multiple regulations, and will lead in demarcating urban growth boundary and come up with new development and trend. This paper puts forward the demarcation methods for urban growth boundary in the new situations, and more technology is worth our further research efforts.

Acknowledgements This paper is technically and financially supported by National Natural Science Foundation of China projects "Research on Tradeoff/Synergy Relationship and Resilient landscape Regulatory Network Mechanism for Guangdong-Hong Kong-Macao Greater Bay Area Water-Land Ecozone (51468064)" and "Dianchi Watershed Urban Space Growth Boundary Research" (51668065); Provincial universities key research of Guangdong projects "Study on rural production-living-ecological spatial allocation and functional zoning red line in the Pearl river Delta water network area" (2020ZDZX1033).

References

Amin, T., & Bryan, P. (2011). Two Rule-based Urban Growth Boundary Models Applied to the Tehran Metropolitan Area Iran. *Applied Geography, 31*, 908–918.

Andres, D., & Elizabeth, P-Z. (1995). Lexicon of the New Urbanism. *Time-saver Standard for Urban Design*, 5–11.

Cai, G. Y., Li, X., Zheng, R. Q. (2017). Study on coupling optimization for establishing urban growth boundary under game theory City. *Planning Review, 42*(03), 19–24.

Chengri, D. (2012). Urban Growth Boundary Theories. *Planners, 28*(03), 5–11.

Claire, J., Scott, G., & Mary, S. (2003). Using the SLEUTH urban growth model to simulate the impacts of future policy scenarios on urban land use in the Baltimore Washington metropolitan area. *Planning and Design, 30*, 251–271.

David, B., Jennifer, F., & Kristen, N. (2004). Public policies for managing urban growth and protecting open space: Policy instruments and lessons learned in the United States. *Landscape and Urban Planning, 69*, 271–286.

Frey, Mary. Urban Growth Boundary [EB/OL]. https://conservationtools.org/guides/48-urban-growth-boundary.

Gerrit, K., & Lewis, H. (2001). The inventory approach to urban growth boundaries. *Journal of the American Planning Association, 67*(3), 314–326.

He, C. Y., Jia, K. J., & Xu, X. L. (2010) Planning method for defining the boundary of the urban and rural construction land expansion based on spatial analysis technology of GIS. *China Land Science, 24*(3).

Huang, H., & Casella, S. (2007). Main policy and cases of smart growth, consideration on how to apply in China Megalopolis Modern. *Urban Research, 5*, 19–28.

Huang, M., Zhang, R., He, Q., Wang, C. (2017). Returning to the origin: Reflections on 'Permanent' and 'Phased' Urban Growth Boundaries. *City Planning Review, 41*(02), 9–26.

Jie, S., Xiaohu, L., Xiaohua, Z., Bin, Y., Yiming, Z. (2016). Urban Development Boundary Delimitation by Six Steps. *Planners, 32*(11), 45–50.

Jing, C., Xingyin, L., Tingting, C., & Zongcai, W. (2016). Anqing urban development boundary specification and management based on spatial growth simulation. *Planners, 32*(06), 23–30.

Knaap, G., & Nelson, A. (1992). The regulated landscape: Lessons on state land use planning from oregon. Lincoln Institute of Land Policy.

Li, H., Meng, F. T., Li, Z. Y., Wang, C. C., Li, P., Li, G. Y. (2014). Yunnan Haba snow mountain ecological safety evaluation based on the landscape ecological pattern. *Applied, 522*, 537–540.

Long, Y., Han, H., & Mao, Q. Z. (2009). Establishing urban growth boundaries using constrained CA. *Acta Geographica Sinica, 8*, 999–1008.

Metro. The region's 50 year strategy for managing growth. [DB/OL]. http://www.oregonmetro.gov/index.cfm/go/by.web/id=29882.

Miao, Z. Z. L. Y. (2010) The patterns of urban containment and its impact to urban development. Modern Urban Research (1), 63.

Minghua, H., Xiaoqing, T. (2008). Reflection on urban growth boundary in the new urban planning. *Planners, 6*, 13–16.

Peter, V., et al. (2002). Land use change modeling at the regional scale: The CLUE-S model. *Environmental Management, 30*, 321–335.

Pijanowski, B. C., Tayyebi, A., Delavar, M. R., & Yazdanpanah, M. J. (2009). Urban expansion simulation using geospatial information system and artificial neural networks. *Int. J. Environ. Res., 3*(4), 493–502.

Richard, S. (1991). Urban growth boundaries. Governor's Office of Planning and Research (California) and Governor's Interagency Council on Growth Management.

Silva, E. A., & Clarke, K. C. (2002). Calibration of the SLEUTH urban growth model for Lisbon and Porto, Portugal. *Computers, Environment and Urban Systems, 26*(6), 525–552.

Wang, Y., Gu, C. (2017). Grid-based spatial evaluation of establishing urban growth boundary: A case study of SUZHOU City City. *Planning Review, 41*(03), 25–30.

Wu, J., & Lu, B. (2014). Dynamic urban model and optimized urban growth boundary with endogenetic land consumption. *Urban Development Studies, 21*(5), 61–65.

Zhang, K., Zhao, Y. L., Fu, Y. C., Zhang, H. (2013). Fractal dimension dynamics and land use in the lake Dianchi Watershed from 1974 to 2008. *Resources Science, 35*(1), 232–239.

Zhou, J. F., Zeng, G. M., Huang, G. H., Li, Z. W., Jiao, S., Tang, L. (2007). The ecological suitability evaluation on urban expansion land based on uncertainties. *Acta Ecologica Sinica, 27*(2), 774–782.

Chapter 2
The Spatial Planning Technology Innovation of the Smart City Under Knowledge Economy and Ecological Sustainability

Le Che, Yinquan Luo, Yangfang Hu, and Sheng Yao

2.1 Core Elements of Smart City Development

In 2008, IBM brought up the notion "Smart Earth", a part of which—the notion "Smart City"- was considered as a tool to tackle with challenges of urban development. The Chinese government devotes great attention to smart city construction, provinces and cities like Guangdong, Beijing, Shanghai and Nanjing have included "smart city" in main projects so as to promote their economic growth. A year later, Guangdong Province was determined as the first site of smart city project, which is together with IBM to create "number Guangdong" and to promote integrated construction of smart city in Pearl River Delta. Existing researches on smart city focus on strategy framework and application of smart technology in various aspects, which is merely a tunnel of realizing narrow smart city, feasible but limited. After all, the final goal that smart city leads to a better life for people in the context of the comprehensive infiltration of knowledge economy, severe ecological situation and rapid urban development is what generalized smart city required.

It is observed by researchers that smart city presents an obvious interaction with knowledge and ecology, (Che & Wu, 2015) as it is shown in Fig. 2.1, and that

L. Che (✉) · Y. Luo · S. Yao
Department of Urban Planning, South China University of Technology, 381 Wushan road, Tianhe District, Guangzhou 510641, China
e-mail: 838583378@qq.com

Y. Luo
e-mail: 506628772@qq.com

S. Yao
e-mail: 151466673@qq.com

Y. Hu
Sustainable Urban Planning, The George Washington University, 2121 Eye Street, Washington DC 20052, USA
e-mail: 361572471@qq.com

© Springer Nature Switzerland AG 2021
W. Li et al. (eds.), *Human-Centered Urban Planning and Design in China: Volume I*,
GeoJournal Library 129, https://doi.org/10.1007/978-3-030-83856-0_2

Fig. 2.1 Interaction between knowledge and ecology

Fig. 2.2 Quantitative correlation analysis of knowledge and eco-city indicators. *Source* drew by authors

The REG Procedure
Model: MODEL1
Multivariate Test 1

Multivariate Statistics and F Approximations					
S=4 M=-0.5 N=5.5					
Statistic	Value	F Value	Num DF	Den DF	Pr > F
Wilks' Lambda	0.02402214	6.03	16	40.353	<.0001
Pillai's Trace	1.80727122	3.30	16	64	0.0003
Hotelling-Lawley Trace	11.31295908	8.61	16	20.615	<.0001
Roy's Greatest Root	8.55249979	34.21	4	16	<.0001
NOTE: F Statistic for Roy's Greatest Root is an upper bound.					

the correlation was verified via SAS multiple statistical quantitative research. In other words, knowledge as well as ecological index are able to generally cover the requirements of smart city, and knowledge characteristics are strongly associated with ecological index (see Fig. 2.2).[1] Based on the correlation, spatial requirements of smart city are explored so as to figure out corresponding spatial planning method, which is a significant topic of smart city.

---

[1] Relativity Saliences of 4 types of verification procedure in SAS multiple statistical analysis are < 0.0001, 0.0003, < 0.0001, < 0.0001. According to international standard, salience is high when Pr > F value is under 5%, which means that the linear relationship between core characteristics index of knowledge city and that of ecology city is highly significant.

2.2   Requirements of Smart City Spatial Development

There are multi-spatial levels in smart city as Fig. 2.3 shows. Carrying physical space, ecological space is an elementary level that exists and is of significance in every period of urban development, and represent the space that obeys the rules of evolution, spatial elements of which include ecological function chunks, tunnel of energy transportation and species, and the natural ecology environment.

Knowledge space is the notion that urban function network becomes innovative with the arrival of knowledge economy era. Also, the development of electronic communication and Cyberspace promotes electronization and virtualization of the network among various function chunks. As an emerging function network level,

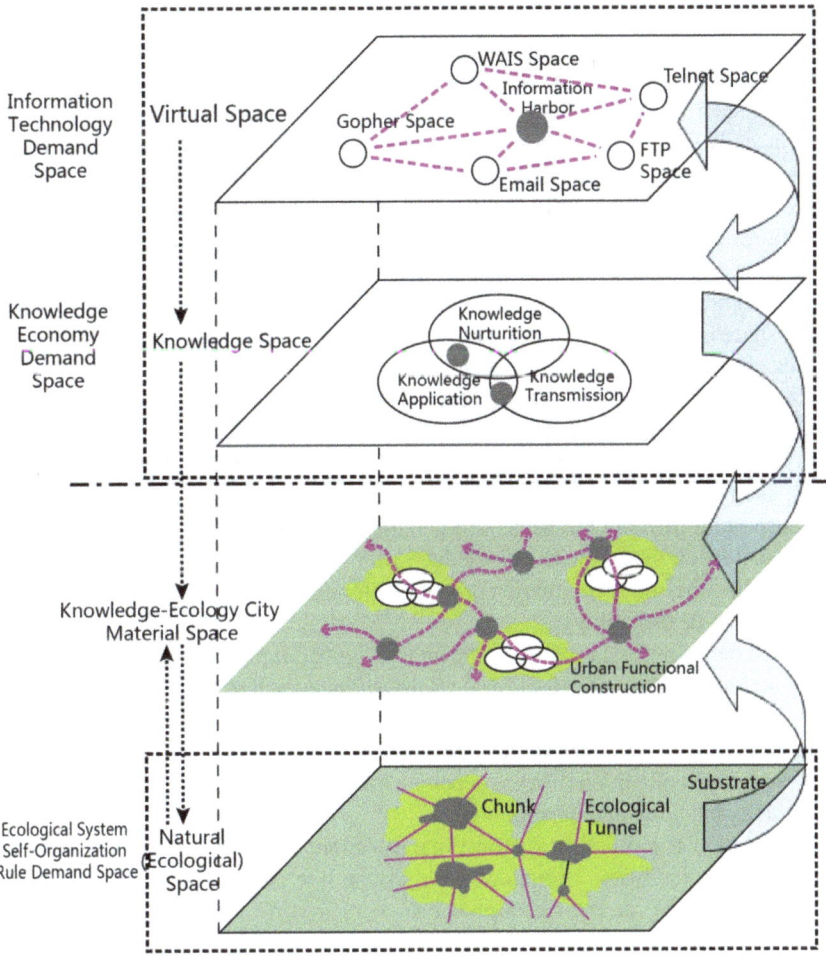

Fig. 2.3 Multi spatial level in smart city

knowledge space largely combines smart space and material space, which presents strong compatibility in spatial characteristics. It is distinctive for certain spatial properties on one hand, on the other hand, it possesses the capability to communicate remotely and to conduct information management as well as knowledge innovation. Thus, knowledge space is probably to generate impacts on original material space with its resilience, flexibility and interactivity, and to trigger revolution of spatial elements, structure and pattern.

Virtual space, consisting of informational flows that portrait material network, is considered to be a personified spatial level with participation of human awareness. To some degree, virtual space is a combination of computer technology and human imagination. Virtual space[2] can also be treated as a conceptual infinite space, consisting of multiple working spaces such as Finger space, Gopher space, Email space, Web space, WAIS space, Telnet space and FTP space, and information transform spaces like information nodes, information harbor, and information exchange center. In knowledge city, virtual space plays an increasingly significant role, which enables people with high-level knowledge to conduct various social and economic activities such as knowledge production, knowledge transaction, knowledge transformation, daily life and recreation activities, and even those real-life cultural activities. Thus, virtual space can also be considered as physical space's projection on cyber space, which thoroughly indicates the complexity of urban space and the pluralism of culture, and stronger interactivity as well as participation, leading to huge impact on urban material space. Urban spatial structure will probably be limited by Internet and broad brand rather than convenience of transportation and availability of land; urban space may be built via software virtually in addition to being built by stone and woods physically. Supported by technology, virtual space is equipped with infinity characteristics that decline certain type of spatial barrier and produce impact on traditional social space bound.

Material space of smart city is discussed at a such multi-dimensional spatial level, and treated as a physical carrier of various economic and social activities. For the purpose of meeting the needs of humanization space, it's necessary to combine ecological factor system with knowledge factor system and virtual spatial system so as to realize the trend of smarter urban development which is consistent with the development of technology.

Based on the analysis so far, we are able to figure out physical space of smart city, the function of which transfers from relatively independent yet connective production, life, and service to knowledge transmission, application and comprehensive living function based on ecological substrate, as Fig. 2.4 shows.

Knowledge cultivation district: accounts for function of educational and training, technology hatching and developing;

Knowledge transmission district: mainly infers to connections like roads and transportation facilities, communication spaces like café and meeting room, and other transmission carriers like high ways;

---

[2] John December. http://www.december.com/.

Fig. 2.4 Physical space of smart city

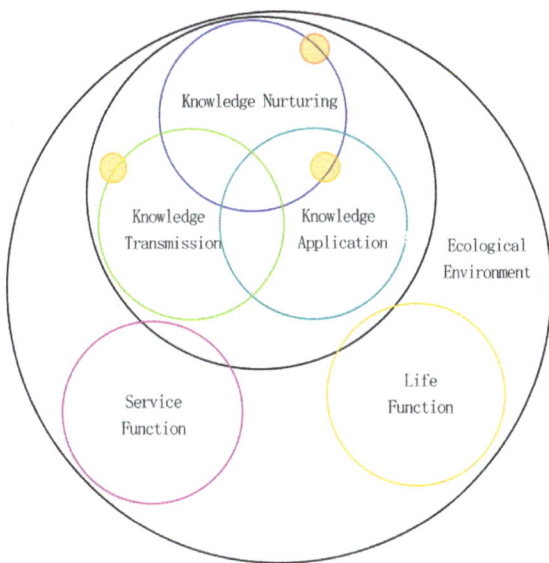

Knowledge application district: equals to industry functional district in modern city, loading district of material economy in knowledge-ecology city, including information technology, innovative industry, biological medicine, modern delivery, and a small part of high-level manufacturing;

Mixed district: mainly for flexible zoning and needs for comprehensively mixed function;

Living function district: mainly for living function, together with living facilities like fundamental education, grocery store and restaurant;

Service function district: includes public living facilities, law, advertising and designing;

Ecological environment district: provides human with function of resources utilization, recreational space and ecological reservation, including green infrastructure and rehabilitation.

2.3 Innovation and Development of Smart City Spatial Development

Given that existing development model of urban spatial is hard to accommodate to the requirement of smart city spatial development, it is necessary to explore land use, spatial structure model, spatial lay out and structure, and spatial shape.

Based on that, the research systematically arranges theory of knowledge city, ecology city, urban spatial development, knowledge management and spatial ecology, one outcome of which is DICE model, providing guidance for the research. The

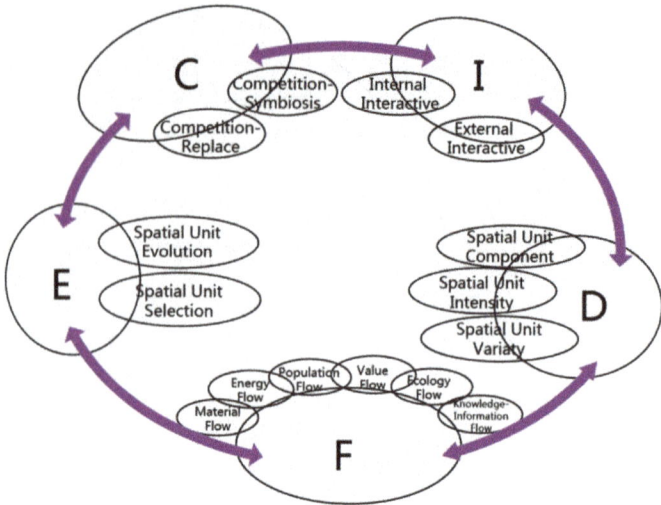

Fig. 2.5 DICEF development model of knowledge and eco-city indicators

core theory of the model is to obey four basic principles of ecology: distribution, Interaction, Competition, Evolution, and the one we added—Fluent, which promotes innovation and development of knowledge and helps realizing development of two core elements of smart city. Referring to the DICEF model (see Fig. 2.5), which is evolved from the DICE model (Zhao, 2015), researchers make effort to figure out a reasonable explanation towards new characteristics that smart city presents, and to guide innovation of planning technology, as Fig. 2.6 shows.

2.3.1 Innovation and Development of Smart City Spatial Development

Spatial evolution has been primarily reflected in of evolution space and zoning, which indicates both the evolution and natural selection process of spatial unit. The result is that 4 spatial level show up in smart city: ecology space, material space, knowledge space and virtual space. With the appearance and evolution of multi-dimensional space, the level of urban function tends to be higher, which facilitates the gathering of knowledge production function, and the appearance of agglomeration of knowledge production function.

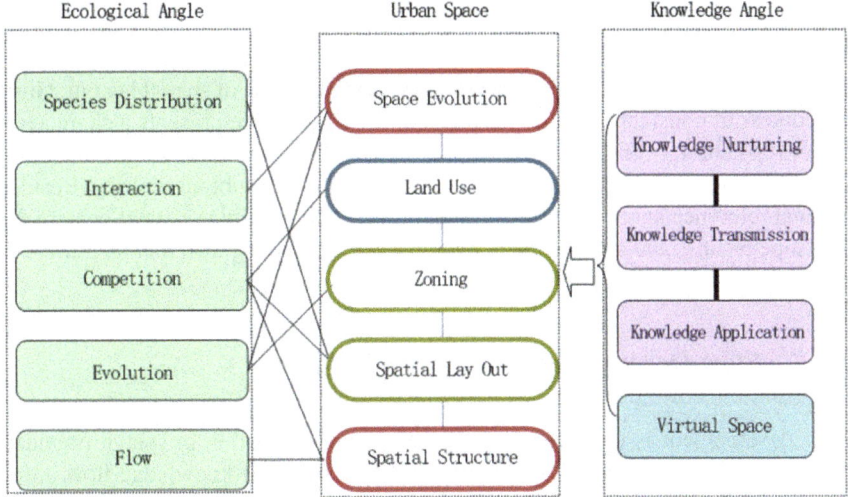

Fig. 2.6 Spatial development framework of knowledge-eco city

2.3.2   *Interactive Development of Smart City Space*

Interactive development of smart city space is indicated in interaction of multi-dimensional space and that of land use, presenting both internal among urban functional constructions and external interaction between them and the environment. The respect of ecological self-organization discipline has become stronger in ecological city than ever before, where natural space and material space interact with each other frequently, impacting and determining the evolution of humanized space together with knowledge space and virtual space. In this process, traditional boundary of these spaces was broken thus knowledge-ecology city presents complicated and vague characteristics, consequently multiple spatial elements tend to be compatible with each other.

2.3.3   *Competitive Development of Smart City Space*

A spatial competition also exists in adjustment of land use and spatial lay out, leading to various possibility of spatial combination, which includes compatible and mixed land use as the result of coordination and symbiosis effects in fusion of working space, residential space, recreational space and production space, and evolution of land use as the result of competitive effects which consist of new urban land use structure model.

2.3.4   Distribution of Smart City Space

The distribution of smart city space refers to transformation of spatial lay out. Given the context of constantly changing spatial units, urban space presents a characteristics of resilient network, fulfilling various functions such as knowledge cultivation, knowledge transmission, and knowledge application as a vibrant entirety. Besides, original reference space pattern has been broken into multiple resilient centers due to compete-substitute and compete-symbiosis effects among different land uses.

2.3.5   Flow Development of Smart City Space

Flow development is considered as an outcome of revolution in spatial organization and structure. Dense flowing network that consists of knowledge flow, information flow and various ecological flow, shows as a framework of urban spatial organization that includes ecological network, knowledge innovation network and material flowing network. Moreover, based on knowledge transmission, the flowing and spreading of knowledge-ecology city turns to infective model, combined with complex transmission characteristics, promoting the transfer of urban spatial structure from the pattern of single center with sphere structure to that of network with multi-blocks.

2.4   Application Practice of New Guangzhou Knowledge City

The study applies DICEF model to the new Guangzhou knowledge city, in the city spatial structure, land structure, land development, mixed ecological open space development, virtual space and physical space and interactive development, to a certain extent to achieve the strategic objectives of the development of smart space.

2.4.1   Evolution of Spatial Functional Organizations

In the smart city, the knowledge industry has gradually occupied the leading position of the industry development, knowledge and innovation have become the most important spatial influence factors (Le & Xiaobing, 2012). In view of this, the new Guangzhou knowledge city spatial organization to promote knowledge and innovation. Focus on the development of knowledge and innovation in the industry, the function of organization planning for the pilot industry, leading industry, industry characteristics, supporting industries, complementary industries in the five level functions, which includes the training of leading industry of science and technology research and development and education, the leading industries including creative industry, high technology, advanced manufacturing industry, including the characteristics of leisure health and fitness recuperation, support industrial and commercial services, including residential resort, complementary industries including culture and sports, business office and Exhibition forum. The selection of specific functions of network is shown in Fig. 2.7. Based on this, the new Guangzhou knowledge city information technology industry, biotechnology industry, new energy and environmental protection industry, advanced manufacturing industry, scientific research, creative industries, financial services, education and training industry as the pillar industry, and the eight pillar industries as the focus, to promote the industry development in depth, to encourage all types of independent research and development activities, as well as industry consolidation activities and subdivision. We will improve the upstream and downstream industry chain of the pillar industry, provide a sound platform for business services and marketing, and promote the development of high-tech industries and producer services. Provide adequate support services, taking into account the various types of living facilities nearby, focusing on Sustainable development.

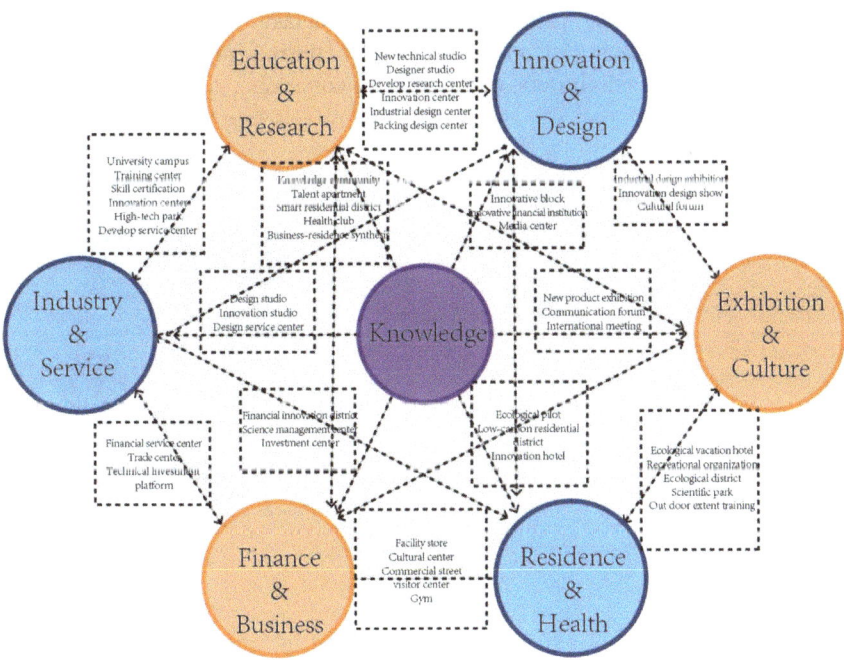

Fig. 2.7 New Guangzhou knowledge city network function selection

2.4.2 Spatial Distribution and Structure: Distribution and Flow Development

The spatial structure planning of the new Guangzhou knowledge city is an elastic multi center network multi group space structure (shown in Fig. 2.8). New knowledge city is located in the northern part of the Guangzhou mountains in the ecological zone. The region is low hill valley terrain, the main part of which is hills with the slope of lower than 25°, large mountains locate on the eastern and western sides, the middle of the terrain was flat and narrow distribution, presenting a more obvious tree shaped valley terrain. The southern part of the land is relatively complete, gentle terrain, the development of land can be more concentrated, with the overall development of the conditions. The western region is the peak of the mountain, possessing the best land-scape and yet the most sensitive ecological environment, which is also an important part of the Guangzhou city green center, where strict control of development is in need.

Based on the comprehensive evaluation of land conditions, managing ecological landscape resources, geomantic environment simulation trend, the continuation of the historical context based style, integrate existing construction conditions, forming a "one lake, seven garden" pattern, Among them, "a lake" as the center of the lake ecosystem and urban living Island, "seven Park" for the red orange yellow green blue purple seven functional clusters, corresponding composite, education, innovation,

Fig. 2.8 Lay out of New
Guangzhou knowledge city

scientific research, industry, service, design seven theme, (Che & Wu, 2013), as shown in Fig. 2.9.

Red: multi compound cluster. Knowledge city is located in the starting area, multiple functions, and focus on the development of knowledge city characteristics.

Orange: education and training cluster. Located in the eastern part of the southern part of the knowledge, with the core of the educational function, with the characteristics of sharing the public center, and combining with the financial investment institution, it becomes the leading area of knowledge production.

Yellow: innovative service cluster. Located in the southern part of the knowledge of the west, with innovative service functions as the core, combined with the west

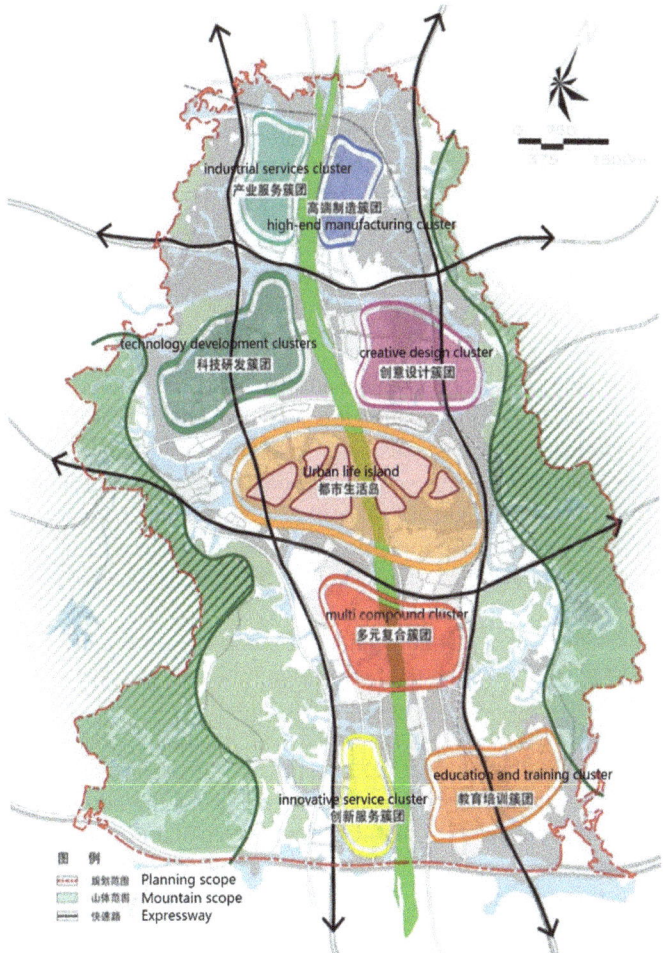

Fig. 2.9 New Guangzhou knowledge city lay out planning (Wu, 2015)

side of the mountain hat ecological landscape resources, leading the city to become the leading areas of knowledge and colorful life.

Green: technology development clusters. Located on the west side of the headquarters of the knowledge city, with the function of scientific research and development as the core, combined with the headquarters of the enterprise and the function of product incubation, it has become the leading area of the science and technology research.

Cyan: industrial services cluster. Located on the west side of the north of the city, taking the industrial service function as the core, combined with the training logistics and other related projects, it has become the leading area of the productive service industry of the knowledge city.

Blue: high-end manufacturing cluster. Located in the eastern part of the northern part of the knowledge, with high-end manufacturing as the core, the knowledge city to complete the transformation of the dominant areas of knowledge.

Purple: creative design cluster. Located in the eastern part of the city of knowledge, creative design as the core, combined with innovative cultural display, highlighting the unique charm of the city of knowledge.

2.4.3   Hybrid Compatible Development: Interaction and Competition

The new Guangzhou knowledge city "mixed land" not only includes the combination of several different functions, reflecting the flexibility and elasticity, but also emphasizes the "suitability", which is considered to the surrounding environment for the land use nature of the constraints on one hand; on the other hand, the compatible relations of leading function of land and other blocks is considered, in the premise of not affecting the dominant land use function, and other functions for mixing the quantitative relationship between the various land use, in order to achieve the coordination and cooperation.

With the development of planning unit (UPD) as the research platform, the control of mixed land can be divided into mixed type control and mixed proportion control. There are two principles for the control of mixed type: one is the principle of land use, the priority of the type of land associated with the hybrid, this hybrid can reduce traffic travel, increase the vitality of the city, improve industrial efficiency. Also there is the principle of no mutual interference, interfering function should not be mixed in the same land. As for mixed ratio control, the proportion of different types of facilities on the ground should be measured. In general, the proportion of dominant land use can not be less than 50%, and the proportion of land mixed with the dominant land shall not be less than 20%.

According to the development characteristics of the new knowledge of the city, in addition to considering the overall structure determine the structure of the green wedge, based on the perspective of ecological protection to determine ecological networks, have more stringent requirements for a class of residential land, planning structure and development direction of the external traffic land, professional strong municipal facilities land and the location strong production protection land, other land property, including land for public facilities, industrial land, residential are can be mixed use land properties. The first planned unit development in accordance with the division of business services, residential, industrial three functions, then according to the leading function of different development unit put forward to adapt its development with mixed mode, and puts forward its development stage and suitable development areas according to the characteristics of each model, determine the quantitative criteria based on mixed degree. In order to guide the land use of the new knowledge city.

2.4.4 Conservation and Development of Natural Ecological Space: Evolution and Competition

The natural ecological space should not only be reflected in the series of good ecological patches, but also reflect the characteristics of the ecological matrix. The new Guangzhou knowledge city to determine the ecological corridor, with green center, and nine water ecological pattern stretches seven peak, take the priority of ecological environment, the integration of regional security system; ecological background, constructing ecological pattern; network hierarchy, improve the eco system; the integration of landscape elements, highlighting the ecological characteristics of landscape development strategy.

According to the principle of ecological priority, we should strengthen the protection and utilization of natural mountain, integrate natural ecological resources, and lay stress on the construction of ecological security pattern. In order to build a safe pattern, promote organic growth as the goal, the natural ecological pattern into the whole project. The use of the base of good ecological resources, with the surrounding mountains as the ecological matrix, the integration of the vertical and horizontal rivers of the base, the formation of the overall ecological pattern. City morphology combined with topography, forming a plurality of wedge channel mountain city, maintain the ecological background and the connection with Maofeng Mountain-Baiyun Mountain green center. To create a central ecological corridor and ecological center of the lake as a land ecological landscape corridor to nucleus, forming a green network, to retain the original ecological base and create a waterfront space. As shown in Fig. 2.10.

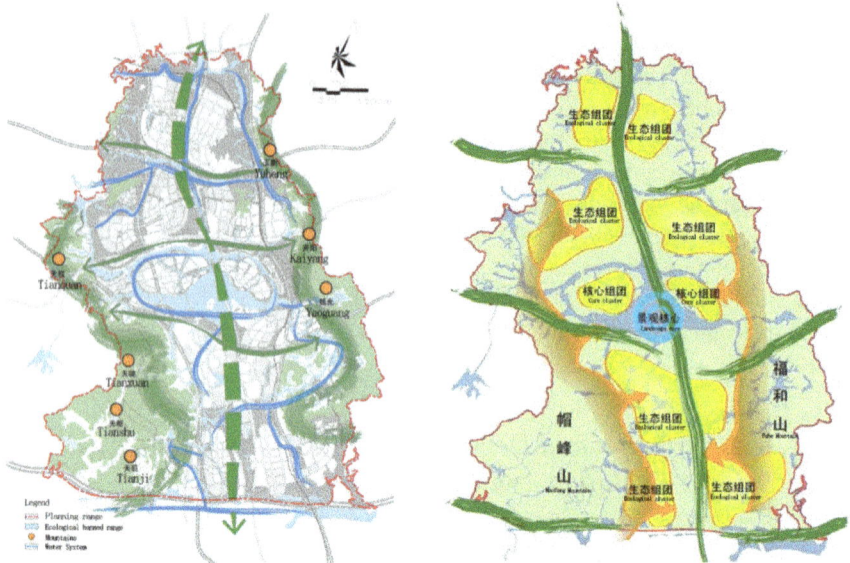

Fig. 2.10 New Guangzhou knowledge city ecological system planning

2.4.5 Interactive Development Between Virtual Space and Physical Space

There is an interactive mode in the new Guangzhou knowledge city that the physical space elements conducts analysis, simulation and optimization via virtual network system that has a high degree of perception, interconnection and intelligent, so that the city can balance all aspects of demand, optimize the allocation of various resources of the city. As shown in Fig. 2.11.

In the interactive framework, the city is perceived, and the virtual space can get the correct perception and measurement of the condition of the important management objects in the system. The city is also interconnected, and has been established for a long time between the decentralized urban system and the urban industry to establish efficient communication and interaction. Besides, the city is intelligent, where collaborative virtual network carries out multi aspect and multi dimension analysis and predictions and optimization through various sources of information to achieve intelligent decision-making of main entity organizations such as enterprises and institutions of the city, etc.

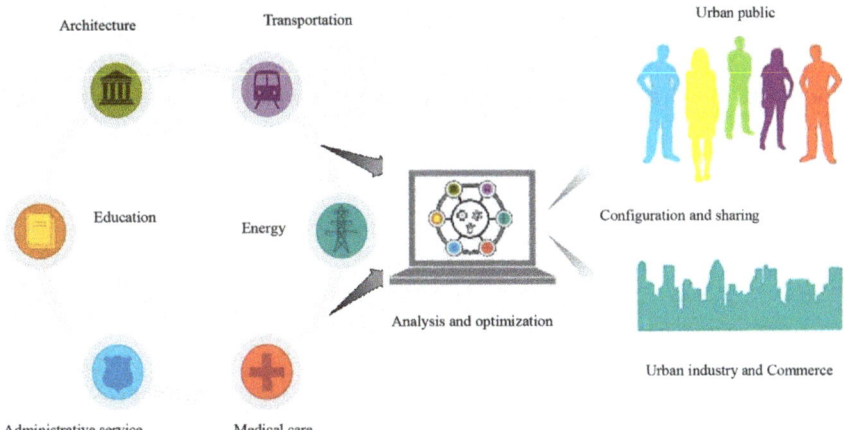

Fig. 2.11 Interaction model between virtual system and physical spatial

2.5   Conclusion

This article interprets the wisdom of the city's core idea from the inspects of information technology, knowledge economy and ecological sustainable development, and conducts a relevance study to verify the interactive circular relationship between knowledge, wisdom, and ecology. As for the demand of the smart city spatial development, the research on information technology, knowledge development, ecological self-organization is in need. Based on this, and the DICE model from knowledge-ecology theory, DICEF model is established to guide the space development of smart city from spatial distribution, spatial competition, spatial interaction, spatial evolution, spatial flow and other aspects, to explore space demand and spatial planning strategy of development trend, and apply this to the practice of Guangzhou Knowledge City, to guide and practice the strategy of smart city in physical space.

Project Supported by the National Natural Science Youth Foundation of China in 2014 (Project title: Research on Spatial Development of Urban New Districts in Pearl River Delta from the Perspective of Correlation between Knowledge and Ecology; Project Number:51408234).

References

Che, L., & Wu, Z. (2013). Knowledge and ecology interactionism on space. South China University of Technology Press.
Che, L., & Wu, Z. (2015) A research on spatial changes under the knowledge economy and ecological sustainability. *Urban Planning Forum*, 20–26.
Feng, Z., Cuangeng, Z., Yong, Z. (2004). New influencing factors of regional spatial structure in information era. *Geography and Geo-Information Science*, 98–103.

Le, C., & Xiaobing, D. (2012). A research on spatial changes under the knowledge economy and ecological sustainability. *Urban Planning Forum, 5,* 44–49.

Wu, Z. Q.(2015). General development plan of China-Singapore Guangzhou knowledge city.

Zhao, W.B., & University, D. (2015). Construction of knowledge ecosystem based on dice. *Information Science.*

Chapter 3
Chinese Traditional Ecological Wisdom and Contemporary Sustainable Landscape Art

Yan Huang

Ecological problems are essentially cultural problems. During the past several millennia where human beings and their living environment coexisted, people had created and experimented with many plain and experimental ecology ideas manifested as simple methods and principles to deal with their surroundings and natural. People from ancient China, however, elevated such plain and experimental ideas to a philosophical and aesthetical level, bringing a special and complete series of ecology wisdom into being. This traditional ecological wisdom played an important role in guiding human behaviors and had proved to be safe and effective over centuries, achieving ecological and aesthetical excellence. This not only provides a lens through which to reflect on today's ecological crisis but also provides resources and inspirations for global ecological theory and experiments.

Generalizing this traditional ecological wisdom will be helpful in dealing with the contemporary ecology crisis. This is an effective way to ensure ecological safety, construct sustainable residential environments, and promote and spread Chinese traditional culture.

3.1 From the Plain and Experimental Ecological Concept to Artistic Expression

The evolution of Chinese ecological ideas started from the primitive fear of nature and evolved into the worship and appreciation of it. When people put themselves into the natural world and tried to make friends with nature, they aroused aesthetic emotions and tried to express them through art. Eventually, these abstracted concepts

Y. Huang (✉)
Tsinghua University, Beijing, China
e-mail: Yyhuang1118@163.com

© Springer Nature Switzerland AG 2021
W. Li et al. (eds.), *Human-Centered Urban Planning and Design in China: Volume I*,
GeoJournal Library 129, https://doi.org/10.1007/978-3-030-83856-0_3

Fig. 3.1 The development from the plain ecological concept to artistic expression

combined with philosophical ideas and brought about the aesthetic re-creation of nature (Fig. 3.1).

The plain ecological concept mainly concerns nature in an experimental, shallow and limited fashion within the boundaries of subjective practical workmanship. It is a program with the lowest cost possible, based on the productive forces, level of technical advancement and current world view, as well as the positive results of the agricultural civilization and the accumulated experiences in interacting with nature. When elevated to a philosophical and cultural level, the concept influences every aspect of society, including art and attitudes towards living.

The traditional Chinese ecological concept differs from that of the west, which is based on the concept of reflecting and making changes after industrial civilization has done drastic damage to the environment; it "is an innately natural attitude that is related to specific production environments, and has complicated relationship with the capability of experiment, value orientation, theoretical concepts and way of thinking."[1] The concept also "declares a special attitude towards nature from the Chinese nation in all of Chinese poetry, art and religion, and expresses that the Chinese philosophy rules with a fearless ideology."[2] The concept has formed a special and effective ecological cultural system, which is important in guiding people's thinking and behaviors.

Almost 300 years of industrial civilization in the west brought environmental damage and ultimately caused a global ecological crisis. The crisis forced people to think and reinitiate the search for how to correctly treat humans and nature. Over time, the ecological concept in traditional Chinese culture became an important intellectual source for global ecological theories and experiments.

> "Westerns have realized the ecology crisis accompanying with the industrial development and should have to explore the harmonious development with nature from eastern." H. Royston, an American ecological ethics scholar purports, "Eastern and western should learn from each other when China has been stepping into modern era."[3]

3.1.1 Source of Traditional Chinese Ecological Wisdom

Traditional Chinese philosophy, such as, "For I am abstracted from the world, the world from nature, nature from the way, and the way from what is beneath abstraction" by Lao Tzu and "The people are my brothers; the creation is part of me" from

---

[1] Li (2006).

[2] Simons (1982).

[3] Royston (2000).

Zhengmeng by Zhang Zai, directly set such aesthetical critique standards as "artificially made but looks natural" and "The nature can carves it pretty" for gardens and crafts as a heavenly and supernatural work. Other theories include "Tao is embedded everywhere in the universe just as all rivers on land would finally meet in the sea"by Laozi, "Heaven and earth have big beauty without words" by Zhuangzi and "The wise are fond of rivers, while the benevolent are fond of mountains" by Confucius.

In many aspects, including city location, layout, planning, garden, and architecture, confucianist thinking like Xunzi's "All things live in harmony, and grow with nourishments" are manifested. This concept values harmony between man and nature, promotes the pursuit for ecological aesthetics, and emphasizes harmony and union between the subject and object.

Relationship between humans and nature has been considered in Chinese civilization since ancient times, from which the concepts of "man is an integral part of nature" identified by Chinese old sayings, was proposed. Taoism supports an attitude similar to "laissez faire", stating that human should respect nature and that human behavior should be subjected to the laws of nature. In this aesthetical experience, subjective emotions of man become one with the all surroundings within the universe.

Formal language reflects the human understanding of the universe and man's dreams and ideologies towards life. People in ancient China, especially humanitarians, mostly appreciate the beauty of simplicity and plainness, and value the originality of nature and find nature in oneself. This aesthetical concept is shown on the ancient utensils, landscape paintings and even more notably in the gardens of southern humanitarians and urban constructions. The traditional philosophy of union between nature and man has not only become an aesthetical and cultural concept, but also an ecological concept.

3.2 Aesthetical Representation of Traditional Chinese Ecological Concepts

3.2.1 Figurative Expression of Ecological Concepts: Utensils

The aesthetical representations of Chinese ecological concepts can be figurative, abstract, or a mix of both. Beauty is the combination of spirit and substance, and the combination of natural, cultural and artistic beauty.

People from the past respect, fear, or even worship the "thing", so they followed the Chinese traditional ideas of "adjusting measures according to local conditions" and "teaching students according to their aptitude", which means to design utensils according to the natural shape or color of the material. It is a way of implementing creativity within the limits of outside conditions. Thus, the aesthetic standards required that arts and crafts be designed and made according to the material's character, be in harmony with the surroundings and make use of the local materials and resources as much as they can.

Fig. 3.2 Jadeite cabbage at
the national palace museum
in Taipei

It took really tough, time consuming labor to make the raw jade into the finenest artwork because the jade is a very valuable material. Therefore, maximizing labor and material utility was the key principle, which led to the "measuring material according to it's texture" idea as its artistic idiosyncrasy. Each raw jade utensil has a unique texture, color and shape that evokes imagination and creative fantasy, and these should be the most important aesthetic ideas of the utensils.

The jadeite cabbage is one of the most precious treasures of the National Palace Museum in Taipei. It is a piece of jadeite carved into the shape of a cabbage, and is very realistic. The craftsmen utilized the original green and white color of the jade to make the leaves and stems of the cabbage. A creative mind and superior carving skills made this a masterpiece (Fig. 3.2).

Another example is the Duanyan ink stone from Guangdong province. The craftsmen not only hammererd or chiseled the ink stone into egg shapes, rectangles, squares and circles according to its natural shape, but also fully utilized the natural tile and combined elements from literature, history, drawing, calligraphy to turn the defects on the ink stone into an impeccable comprehensive artwork (Fig. 3.3).

3.3 Two Abstract Expressions of Ecological Concepts—Landscape Painting and Poetry

China is a mountainous country, where two thirds of the total area are mountains. Archeological findings prove that Chinese primitive and agricultural civilizations

Fig. 3.3 Guanddong Duan Yan ink stone at Guangdong Museum

mostly developed in foothills or hillsides of mountains. The important role of mountains in Chinese culture was determined by Chinese natural and geographical features. Water worship appeared very early in history as well, and was more involved in aesthetic and philosophical cultural elements. During this time, painting also achieved a breakthrough due to popular appreciation for landscape and landscape poetry.

The only painting by Zhan Ziqian (Sui Dynasty, about 550–604) that survives today is "Strolling About in Spring," which is a perspective arrangement of mountains. It has been cited as not only the earliest surviving work of Chinese landscape painting or the first shan-shui painting, but also the oldest existing Chinese painting (Fig. 3.4).

Different from the utensils which imitate nature and represent the ecological concepts in a figurative way, Chinese traditional landscape painting has deep

Fig. 3.4 Trip in Spring, Zhan Ziqian (about 581–618 A.D., Sui Dynasty), at the National Palace Museum in Beijing

Fig. 3.5 Suzhou museum, designed by Leoh Ming Pei. Photo taken by author

spatial ideas and philosophical meanings, presenting ancient people's environmental concepts, which were abstractions representing the ecological ideas and which also have similarities with contemporary art (Fig. 3.5).

The landscape painting is the inevitable sequence of the admiration of nature, the awakening of the personal awareness and the aesthetic perception of the landscape which built the Chinese traditional aesthetic concepts such as "no border between people and nature" and "man is the integral part of the nature", noted by Zhuangzi. Ever since Kaizhi Gu (348–409) created Lushan, landscape painting came into formal existence, possessing the special representation of the landscape culture.

Landscape painting values the simultaneous action of viewing paintings and making paintings to develop deeper aesthetic activities. The first is to return to and become one with the nature and the second is to cultivate sentiments and purify the soul through reality, godly and magical nature to reach a sublime spiritual state (Fig. 3.6).

Additionaly, landscape painting shows an ideal aesthetic realm and lifestyle, just as Xie Lingyun (385–433 A. D., South & North Dynasty) wrote,"*Dusk and dawn appear apparently variant. Hill and stream present distinctively brilliant.*"Tang Yan drew a dream land of the everybody, which meant villa and pavilion in the mountains surrounded by water, along with poems, painting and music in which to express landscape reverence and worship (Fig. 3.7).

Fig. 3.6 Light snow over a fishing village, partial, Wang Xi (1037-? A.D., Song Dynasty), at the national palace museum in Beijing

Fig. 3.7 Tea painting, partial, Tang Yan (1470–1523 A.D., Ming Dynasty), at the national palace museum in Beijing

3.4 Producing and Artistic Recreating—Traditional Chinese Ecological Contribution and Manifestation on Human-Built Environments

Traditional Chinese ecological concepts are also reflected on ancient urban planning, site selection, landscape and architectural design. The living environment for humans, from its form to functioning organization, all run in the same groove as nature. There is evidence of which in the recycling of different materials and energies, automatic recovery from natural disasters, and the mutualism among different elements. These form a unique ecological concept and are realized in their contributions to the environment in gardens and urban constructions, resulting in not only a special natural aesthetic effect but also a minimization of workload, earthwork, as well as expenses on daily conservation, management and operation work.

Chinese traditional gardens appreciated dealing with the landform naturally and smoothly, maximizing the utility of the original topographic conditions, building the terrace, hill and pond with the least amount of earthwork. The river was located along the hillside and both elements depended on each other. The river obtains the aesthetic, spatial and ecological functions through its natural character instead of its volume or area; all manmade water systems will surely imitate this natural form.

For instance, a Chinese proverb "water flows down to the lower places", which describes the natural movement of the water and evoked many forms of water included falling, springs, torrents, brooks, lakes and wetlands. It will be turned out the special natural aesthetic effects with least amount of labor, earthwork, daily maintenance, management and consumption (Fig. 3.8). All must learn from nature, and

Fig. 3.8 The master of the nets garden, Suzhou, from the book of Suzhou garden

Fig. 3.9 The master of the nets garden, Suzhou. Photo taken by author

there will never be western fountains from which the water will be sprayed upward in opposite direction of water's fall by mechanic power.

Substances in the garden are fully utilized. When there is no other way, workers dig up earth from the pond and use it at the stop; forming it into a hill is the most common method. The treatments on the pond shore are mostly made of rocks because the south is often rainy and humid and earth shores easily collapse during rainy season. The way the rocks are stacked is an imitation of how varied in height they look in nature and also protects the water (Fig. 3.9).

Abandoning oneself to nature is the principle of city planning. It means abiding by natural rules and taking advantage of natural resources. Just as Guanzi said, "All sites for cities must be at foot of a grandiose mountain or beside the upper reach of a spacious river", an idea which identifies factors considered in determining city location but also the natural ecological laws of drainage and waterlogging as well.

In terms of urban planning, water body construction and natural water systems are skillfully connected, giving them not only functions like military defense and water transport, but also generally useful functions such as water supply, waterproofing, flood control and drainage, and agricultural irrigation, while contributing to climate regulation and beautifying the urban environment (Fig. 3.10).

The aesthetical realm dictated by Chinese traditional ecological philosophy influenced people's lifestyle, values and taste. Tao Yuanming wrote, "When picking asters beneath the Eastern fence, My gaze upon the Southern mountain rests". Lanting Studio, Painted by Wen Zhengming (Ming Dynasty, 1368–1644) described Wang Xizhi, Xie An and some other friends chatting and communicating in Lanting (Fig. 3.11).

Fig. 3.10 "Along the river during the Qingming Festival"copy, partial, at the national palace museum in Beijing

Fig. 3.11 Lanting Studio, Wen Zhengming (Ming Dunasty), at the National Palace Museum in Taipei

3.5 Application of Traditional Chinese Ecological Wisdom in Contemporary Sustainable Landscape Art

Geomorphology and cultural features lead to historical landscape city ideas and formations. How can we use these traditional sustainable ecological principles in a newly started project? How can landscape design or facilities harmonize with the surrounding environment and residents instead of merely wasting the current resource system? How can we utilize existing resources instead of high technology to build sustainable environments in places lacking ecological regulations? The key problem here is to figure out the relationship between landscape and nature as well as the relationship between landscape and people.

To be specific, we want to know how to embody the abstract traditional ecological wisdom through figurative contemporary landscape art. For example, how can landscapes make maximum use of surrounding resources and adopt the simplest and most effective methods, as well as how to influence people's attitudes and daily behavior ecologically through environmental design in order to make sustainable ecological concepts a basic principle of action for people and create a healthy, beautiful and friendly environment.

The landscape practices below could help explain how to apply traditional Chinese ecological concepts to practices and contribute to sustainable development.

3.5.1 Case 1: Heritage Park as a Living Body: Landscape Design of Jingdezhen Imperial Kiln Ruins Park

According to the Chinese utensil's utility of the material's character and maximizing the use of them, designers managed the site's original terrain and saw it as a sculpture created by time and people through local accumulation, digging holes, smoothing and grooving. In the sculptural form, both the plants and the material are recycled within the site and are constantly broken down and reconstructed following the growth of the plants and the erosion of the particles. Discarded rocks from site developments and surrounding building renovations are regrouped and reused. This type of recycling plants and material on site creates an ecological circulatory system in both time and space—the material and visual connection evokes a sculptural sense in the landscape elements (Fig. 3.12).

The physical environmental process has been changing and sculpting the landscape with attainable resources continuously by comprehending how the materials, plants, earth and water interact with each other accurately, scientifically and ecologically. "Indeed, it's the key step from the perceptual, instinct and primitive ecological experience transferring to the rational and general ecological consciousness."[4]

At the source of the ceramic river at the park, the stone surface is perfectly sculpted. Its sloped angle not only makes it look like its perspective line disappears in the sky, but also reflects the sunlight to the greatest extent, thus demonstrating the purpose of environmental aesthetics. Landscape of the earth and sky form dramatic expressions; abstracted combinations of different elements and their sequences create a comprehensive landscape experience (Fig. 3.13). This is the figurative and physical expression of the Chinese traditional cosmology and space concept, similar to the relationship between landscape paintings and Penglai Fantasyland.

The idea of the "ceramic river" in Fig. 3.14 comes from the goal of recycling material through collecting and utilizing rain water on site. In addition, the form of the "ceramic river" is inspired by the Chinese idiom "a distant source and a long stream", which literally means the source is far away and the water flows for a long

---

[4] Huang (2016).

c. 场地气质

Fig. 3.12 Ceramic River: reusing the waste ceramic to form a river that can collect the rainwater in the site. Design drawing: Guo Yijia

distance, and figuratively means something has a rich history and deep roots, just like the long history of Chinese ceramics. Therefore, the "ceramic river" naturally and culturally expresses the concept of ecology.

3.5.2   Case 2: Culturalized Ecological Reservation Zone—Landscape Plane of Nihewan National Park

"Instead of dealing with the messy difficulties and principles of ecological design it seemed that we would rather 'play' with ecology through landform, sought to release us from our preconceived ideas about ecology by introducing the idea of the ecology of the traditional art, thinking about ecology in terms of the interstitial spaces of cultural phenomena and emphasizing the relationship between biodiversity and local villagers life quality."[5] This ecological challenge is about new existential challenges in what appears to be the reconciliation of nature and local economic life.

"Minimum human intervention" and "maximum utilization of current condition" are the two critical principles of re-thinking of nature coexisting with humans. Through stimulating all kinds of static and dynamic resources, manifesting the

---

[5] Huang (2016).

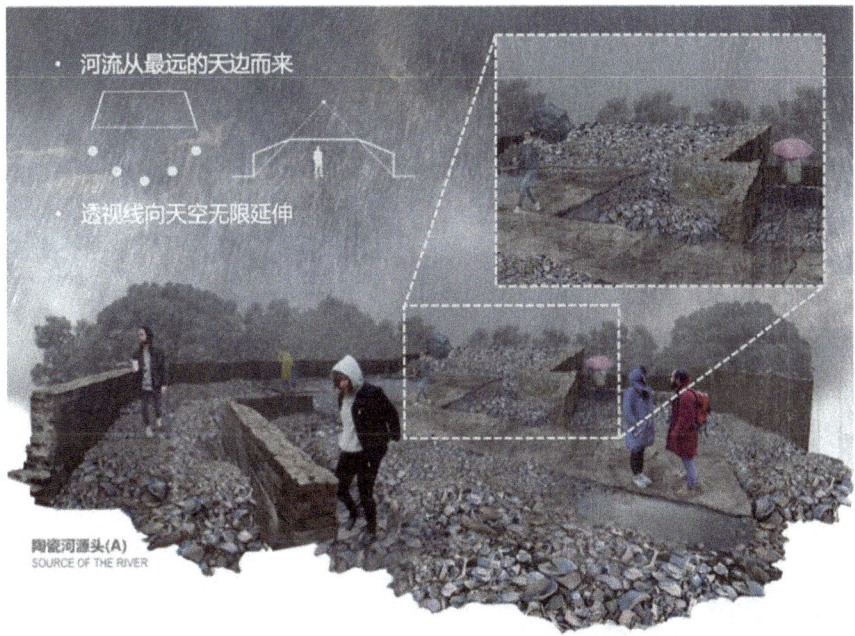

Fig. 3.13 The perspective is endless and stretched into the sky. Design drawing: Guo Yijia

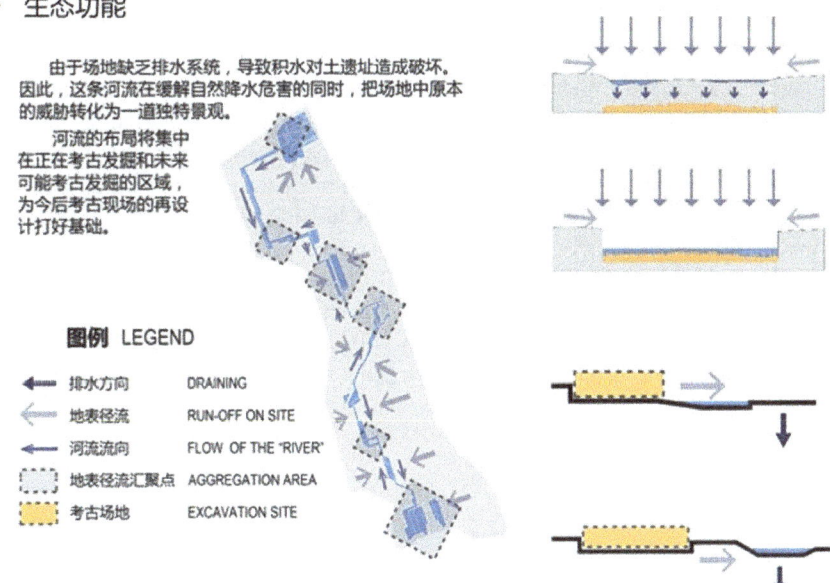

Fig. 3.14 Ecological functions of the ceramic river, Design drawing: Guo Yijia

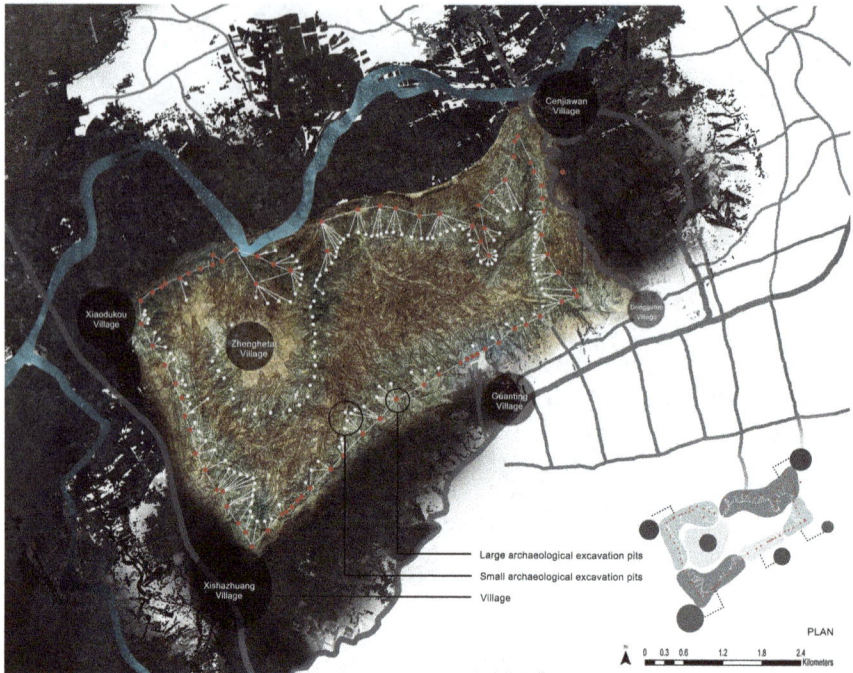

Fig. 3.15 Connecting the rainwater collecting points to form a walking path like a necklace. Design drawing: Ouyang Shiqi, Chu Jingran, Gao Qingyue

villagers' subjectivity and creating a cultural ecology, we can rebuild the rural lifestyle. These are not only attitudes about life and ecological aesthetics, but also idyllic settings, which came from ancient landscape paintings (Fig. 3.15).

Along mountains and terrain, the combination of the walking plank path and the rainwater collection infrastructure along the hillside and landform transports the rainwater to the villages below the mountains. Every single village along the path is connected with it. It's also an ecological community along which the drought plants are refreshed, the earth is re-fertilized by sheep manure and the sheepherders are paid. In this manner, a linear system will be constructed combing rainwater collection systems, grazing areas, walking paths and rural landscapes (Fig. 3.16).

3.6 Conclusion

With the strengthening and deepening of ecological concepts, we can search for ecological ideas, technologies and methods from Chinese traditional art, craft and lifestyles. Besides the application of the visual form, we have to explore inner ecological ideas and build a connection between form, concept and their long-term development when we try to apply these artistic methods to the contemporary environment and landscape construction.

Fig. 3.16 A linear system combined with rainwater collection system, grazing area, walking path and rural landscape. Design drawing: Ouyang Shiqi, Chu Jingran, Gao Qingyue

As plain ecological concepts, Chinese traditional art and crafts were low cost and low technology, which were easy to use broadly. Meanwhile, such ecological concepts are the culture and philosophy as well. Thus, the application of traditional artistic ecological ideas to the contemporary environment is a practical and effective way to spread the traditional culture.

From the plain ecological consciousness and the initial ecological enlightenment, people living through ecological crisis in industrial times must try to transfer these ecological concepts into intense ecological consciousness. Then, we will benefit from whatever mankind's model and scenario for living will become.

References

Huang, Y. (2016). *To resolve ecological issue in national park from a cultural view* (pp. 148–155). China Construction Press.

Huang, Y. (2016). The aesthetical landscape representation of ecological ideas, China living collected papers. *China Water & Power Press*, 65–70.

Li, Z. (2016). *The aesthetical analysis on urban landscape based on contemporary ecological concepts.* Tianjin University Press.

Royston, H. (2000). *Environmental ethics*. China Social Science Press.
Simons, J. O. (1982). *Landscape Architecture*. Taibei: Tailong Book Store.

Chapter 4
Spatial–Temporal Variations of Green Space in Metropolitan Area: The Case of Wuhan, China

Zhiyong Wang, Chun Li, and Tixing Yang

4.1 Introduction

The world has entered a stage of rapid urbanization. More than 50% of the world's population lives in urban areas. By 2025, 65% of the population is projected to live in urban areas. "Urbanization has become the dominant human settlement pattern" (Kelly, 1999; Rees, 1997). In China, the basic national policy of "opening up to the outside world" implemented in 1978 enabled China's economy to grow. A series of land reform laws and regulations launched in 1987 accelerated the process of urbanization in China (Cheng & Masser, 2003). At this stage, China has entered rapid urbanization. The main development features of rapid urbanization are: rapid expansion of the city scale, sustained rapid growth of numbers, and mainly rely on the development of industry (Baojun & Peng, 2015). With the fast urbanization process and the continuous expansion of urban built-up areas, the area of urban green space continues to shrink, and has led to a hidden ecological crisis (Urs et al., 2001). If similar urban growth patterns continue, then Chinese cities will see very detrimental impacts. In the transition period from industrial civilization to post-industrial civilization, the relationship between humans and nature has reached a critical turning point (Naveh, 2000).[1]

Z. Wang · T. Yang
Huazhong University of Science and Technology, Luoyu Road 1037, Wuhan PC 430074, Hubei, China
e-mail: wangzhiyong@hust.edu.cn

T. Yang
e-mail: m201773339@hust.edu.cn

C. Li (✉)
Tongji University, Siping Road 1239, Shanghai PC 200092, Shanghai, China
e-mail: lichun@tongji.edu.cn

[1] According to Northam's urbanization curve, the urban population goes from 25% to 70% is recognized as the period of "urbanization accelerates" (NorthamR M 1975).

© Springer Nature Switzerland AG 2021   61
W. Li et al. (eds.), *Human-Centered Urban Planning and Design in China: Volume I*,
GeoJournal Library 129, https://doi.org/10.1007/978-3-030-83856-0_4

Urban green space is an essential part of the urban ecosystem and has unique ecological and social functions (Grove et al., 2006). Green space can alleviate the environmental degradation caused by rapid urbanization, such as mitigating the urban heat island effect (Onishi et al., 2010), regulating air humidity and temperature (Hao et al., 2012; Sun et al., 2012), inhibiting soil erosion (Nagase & Dunnett, 2012), maintaining biodiversity (Attwell et al., 2000), and providing recreation and leisure places (Kong & Nakagoshi, 2006). However, the current continuous expansion of urban built-up areas has occupied a larger range of the original green space (Weber, 2004). The specific negative impacts are the division of large ecological sources into small environmental patches (Dickman, 1987), and the further invasion, isolation, and replacement of low ecologically sensitive pieces (Weber, 2004; Fernándezjuricic & Jokimäki, 2001).

In response to the challenge of the disappearance of green spaces in cities, Chinese cities have successively carried out quantitative research on urban green space patterns and green space system planning practices in the twenty-first century. The quantitative study of the green space pattern primarily focuses on the interpretation of remote sensing images. The Fragstats landscape statistical model, network analysis model, and gradient analysis method are used to analyze the level of urban green space research, heterogeneity, connectivity and differences, and other characteristics (Tao et al., 2013). Also, the quantitative analysis of landscape patterns based on Geographic Information System in Shanghai introduced a basic ecological network system planning (Lang, 2012); Chang-Zhu-Tan put forward the "Green Heart" of urban agglomerations to protect green space from the perspective of the region (Fenghua, 2010); Chongqing used multiple departments. The joint management model has gradually established a dynamic monitoring mechanism for green space (Lanyun & Facheng, 2013); Beijing has adopted a more general consideration of environmental capacity and bearing capacity to ensure the sustainable development of Beijing metropolitan area (Xiaodong et al. 2017). However, the traditional ecological sensitivity analysis based on landscape ecology neglects the development demands of urban population and industry. This method results in a disconnect between planning and reality (Juan et al., 2014); the macro-line marking has caused the neglect of landscape quality, and the drawings " Green Ring" and "Green Wedge" are only supposed nature (Lin, 2015). If we cannot fully understand the reasons and driving mechanisms for changes in the green space pattern in the process of urbanization, then we must consider the responses and policies needed to positively impact urban green spaces and coordinate mutual success for both industry production and living means. We currently lack the planning information to construct and optimize green spaces in urban environments.

When urban development in China entered a new normal state of deceleration and transformation (Baojun & Peng, 2015), the eastern coastal areas and major cities in the central and western regions were in different stages of development (Juan et al., 2014). The large cities in the central region are facing dual pressure to seek development and environmental protection. In this study, Wuhan was selected for analysis because it is an economic, political, and cultural center in central China and presents similar problems and challenges as large cities in the central and western regions. As the provincial capital of Hubei Province and China's geographical center, its rapid urbanization has caused a huge loss in green space. The municipal government has implemented a number of urban greening policies to protect its green space. This study analyzes the temporal and spatial dynamics of Wuhan green space from the period of 1995–2015 through remote sensing interpretation of classified green areas. This study raises the following three questions: (1) How does green space respond to rapid urbanization and change? (2) What are the differences in the characteristics of changes in green areas in different layers and in different directions? (3) How does the change in green space correspond to the implementation of policies and plans?

4.2 Study Area

Wuhan is the capital of Hubei province in central China and is located in the eastern part of the Jianghan Plain, between 113°41′ and 115°05′E and north latitude 29°58′–31°22′E (Fig. 1a and b). Wuhan is divided into three blocks by the Yangtze River and the Han River with near-west-trending belt-shaped hills in the south of the city. Many lakes are embedded on both sides of the river, forming a complex terrain where water systems and mountains interweave. Wuhan is also known as the "The city leading to the Quartet" and "The city of hundreds lakes". As of 2015, Wuhan has a total area of 8569 km², a permanent population of 10.6 million, an urbanization rate of 79.4%,

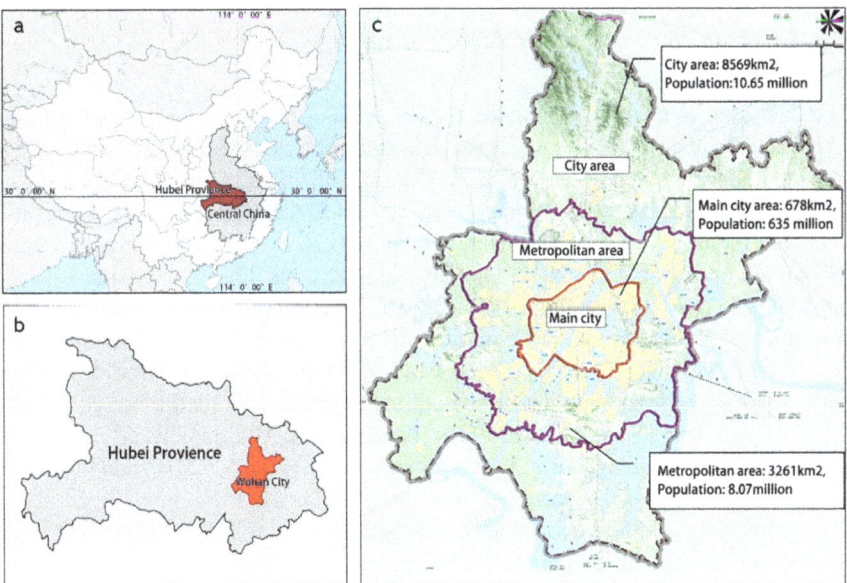

Fig. 4.1 **a** Location of Hubei Province about China; **b** Location of Wuhan about Hubei Province; **c** Location of the study area, Wuhan metropolitan area, in Wuhan city

and urban construction land of 866 km^2. Wuhan is located in the center of China, and it is also one of the most representative big cities in central China. Since the national policy reform to open up to the outside world, the pace of development in the central region has lagged behind the eastern coast. The introduction of the "Rise of the Central China" national strategy in 2004 and the subsequent opportunity for industrial transfer in the eastern coastal areas have given the city new opportunities for development.

Urban space is mainly composed of green space and grey space. Grey space refers to impervious lands such as city buildings, roads, and parking lots (Rees, 1997). This study focuses on greenness consisting of urban forests, grasslands, and farmland and quantitatively analyzes the spatial–temporal dynamic changes of urban green space area, composition type, and development direction during urbanization. The area of this study is 3261 km^2 (Fig. 1c), including the main urban area and parts of the suburbs. The process of urbanization in Wuhan is in line with the concentric circle model, and it expands radially outwards from the intersection of the Yangtze River and the Han River. From 2012, in response to the ecological threat brought about by the rapid development of the city, Wuhan has issued a series of plans such as the "1:2000 Basic Ecological Control Line Plan for Wuhan Metropolitan Area". Wuhan also launched some "Green Hearts" and "Greenway" construction.

4.3 Material and Data

4.3.1 Data Sources and Data Processing

The data used in this study included remote sensing images, Wuhan's overall plan (2010–2020) and Wuhan's land lease data interpretation of the following remote sensing images with no or a few clouds: December 5, 1995, April 20, 2005, Landsat 4–5 Thematic Mapper (TM) image acquired on December 30, 2010; Landsat 7 enhanced theme map plus (ETM+) image acquired on February 26, 2000 and Landsat 8 obtained on October 25, 2015. Operational where the TM image (band 1–5 and band7) has a spatial resolution of 30 m; the ETM+ image (band 1–7) has a spatial resolution of 30 m and a spatial resolution of 15 m in the micro full-color band (band 8); OLI-TIRS images (bands 1–7 and band 9) have a spatial resolution of 30 m, and a thermal infrared band with a spatial resolution of 15 m in the full-color band (band 8) and a spatial resolution of 100 m (band 10–11). Additionally, Wuhan's basic ecological control line plan for 2010, Wuhan's overall plan for 2010, and land lease data for 2014 were used to interpret the coordinated images.

The analysis of the five satellite images involves the following steps (Fig. 4.2): (1) Pre-processing of the picture including geometric correction, atmospheric correction, and metropolitan area clipping. Since Geospatial Data Cloud site provided the dataset of this study, Computer Network Information Center, Chinese Academy of Sciences (http://www.gscloud.cn), it has undergone system radiation correction, and

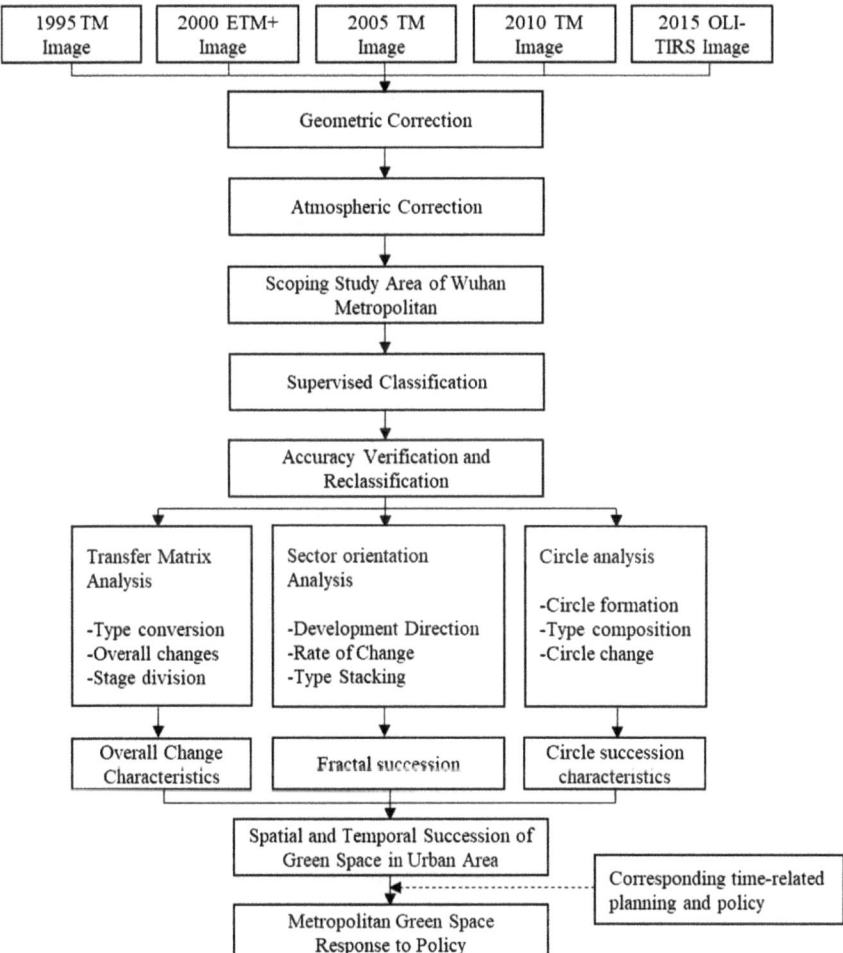

Fig. 4.2 Flow chart of the analysis

geometric correction of map control points and the accuracy meets the requirements of supervised classification. Wuhan's master plan is used to determine the scope of Wuhan metropolitan area to tailor the image; (2) Interpretation of the picture, including the process of band rendering, image classification, and post-classification processing. The band rendering in this study uses bands 5, 4, and 3 for standard false color synthesis (CIR), which is commonly used for monitoring different vegetation types. Through the selection of feature samples and the processing of the maximum likelihood classifier, the images are initially classified. Then, the majority/minority and clump analysis were used to eliminate broken small patches. (3) Verification of classification results. This process uses Wuhan land lease data to verify the classification results. The results show that the Kappa coefficient is between 86.3 and 95.8%,

Table 4.1 Classification of urban land use types

Urban space type		Land cover type	Typical characteristics of land use types
Urban green space	Greenness	Forest	Forest land for growing deciduous forest, evergreen forest, mixed forest etc.
		Farmland	Land for cultivating crops, including paddy fields, dry land and vegetable plots
		Grasland	Forest lands, shrub lands, woodland and garden lands etc.; grasslands, including shrub grassland, open forest grassland and artificial grassland
	Water	Water	Natural land waters and water conservancy facilities include rivers (aqueducts), lakes, reservoirs, ponds and beaces
Urban grey space		Wasteland	Hard-to-use land or vegetation with <5% coverage of vegetation includes sandy land, saline land, bare land, and bare rock
		Build-up land	Urban and rural residential areas, as well as industrial and mining, transportation land etc., including urban construction land, rural residential areas, factories and mines, industrial parks, and transportation roads, but does not include urban green space

which meets the image interpretation requirements and can be used for research (Chao et al., 2009).

4.3.2 Land Use Classification System

The classification of urban land cover types is the basis for remote sensing image information extraction and analysis of green spatial pattern. According to the data classification system of land cover natural attributes and green space surveys in Wuhan Metropolitan Area, the interpretation of remote sensing images is classified into forests (leaves, evergreens and mixed forests), grasslands (gardens, tidal flats, parks with low tree densities, and golf courses), farmland (paddy fields, dry fields and vegetable plots), wasteland (exposed open-air rocks and abandoned land), water areas (river, lakes and reservoirs) and built-up land (impervious to water, including residential, commercial and industrial estate, and public transport use areas). Category 6 (Table 4.1), where forest, grassland, and farmland were used as the landscape composition of the green space in this study.[2]

---

[2] According to the hydrological data of Wuhan, there is a significant amount of hydrological difference between the summer wet season in Wuhan and the rivers and lakes in the dry season in winter and spring, and the long-span distribution of the remote sensing images in this study is within one

4.3.3 Indexes and Visualization Method

To identify the temporal and spatial evolution of the green space in urban areas, we propose three methods of analysis. The first method is a quantitative analysis of the overall green space classification based on the transfer matrix and dynamics. The transfer matrix is used to reflect the conversion relationship between one type of land and another type of area. The rows in the land-use transfer matrix (Table 4.2) indicate land-use types at time T1, and the columns show land-use types at time T2. Pij represents the percentage of total land area converted from land type i to land type j during T1-T2 (Rui, 2010). The degree of dynamics (K) reflects the degree and rate of change of a specific landscape type area within a specified period. In the formula $K = \frac{Ub - Ua}{Ub} \frac{1}{T} \times 100\%$, Ua and Ub are the beginning and end of the study period, respectively. Ua and Ub are the area (m^2) of this type of landscape at the beginning and end of the period, T is time. (Zhang Biao & Jie, 2016).

The second method is to analyze the land composition of each layer by circle analysis. The circle analysis method adopts the graphic fractal technology and uses a circle with a radius of 5 km centered on the geometric center of the Wuhan metropolitan area. Then based on this initial circle, five additional circles are expanded at 5 km increments. These five circles include the main area of Wuhan metropolitan area. Finally, a circle is extended outwards at a distance of 12 km, including all the land outside the main body (Fig. 4.3). Based on this, statistics and analysis can be performed on various types of data in each circle. For example, to study changes in land use at different distances from the center point, or to analyze changes in the distances between the same site and different time periods and specific center points. This can then yield a certain spatial distribution.

The third method is a sectoral fractal analysis of urban erosion characteristics. According to the actual situation of urban development in Wuhan, this study selects the point of intersection of the Yangtze River and the Hanjiang River as the origin point. This point is also primarily located in the geometric center of the

Table 4.2 A sample of the land use transition matrix (Rui & Dao-Lin, 2010)

		T2				P_{i+}	Reduction
		A_1	A2	…	A_n		
T1	A_1	P_{11}	P_{12}	…	P_{1n}	P_{1+}	$P_{1+} - P_{11}$
	A_2	P_{21}	P_{22}	…	P_{2n}	P_{2+}	$P_{2+} - P_{22}$
	⋮	⋮	⋮	⋮	⋮	⋮	⋮
	A_n	P_{n1}	P_{n2}	…	P_{nn}	P_{n+}	$P_{n+} - P_{nn}$
	P_{+j}	P_{+1}	P_{+2}	…	P_{+n}	1	
Addition		$P_{+1} - P_{11}$	$P_{+2} - P_{22}$	…	$P_{+n} - P_{nn}$		

year. Therefore, although the water area belongs to the green space, it is not used as a quantitative evaluation value in this study, only a reference value.

Fig. 4.3 Diagram of circular
analysis

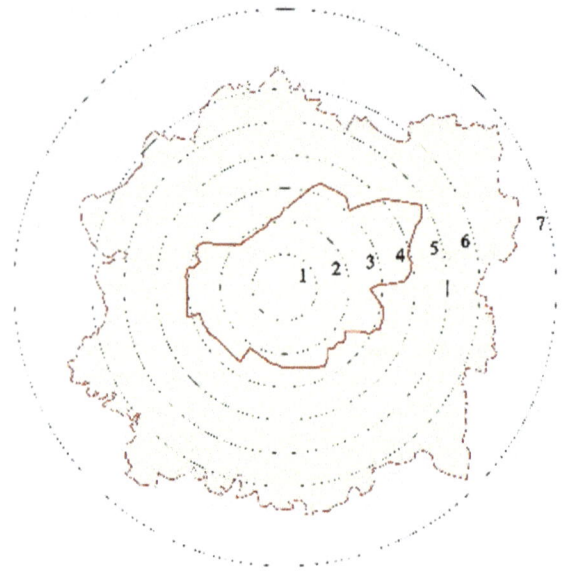

Wuhan metropolitan area and uses the origin point as the sector in 8 directions. The corresponding instructions of each segment are as follows: I = northwest, II = north, III = northeast, IV = east, V = southeast, VI = south, VII = southwest, VIII = west (Fig. 4.4). This information and the use of the statistics of the green space changes in each sector can be used as the basis for determining the different erosion directions.

Fig. 4.4 Diagram of sector
analysis

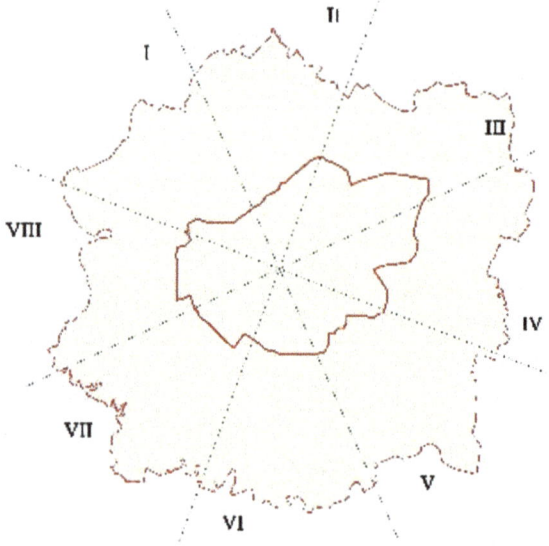

4.4   Results

4.4.1   Evolution of the Quantity of Green Space

During the 20 years from 1995 to 2015, the land use pattern in the Wuhan metropolitan area has undergone tremendous changes (Fig. 4.5). The rapid expansion of the built-up area has led to the massive loss of green space in the metropolitan area, and the urban green space rate has dropped from 75.13 to 53.35% (Table 4.3). This study summarizes the overall and categorical changes in the green space through the transfer matrix and dynamic rate of change.

According to the data provided by the transfer matrix (Table 4.4), the percentage of the built-up regions within the Wuhan Metropolitan Area increased from 8.85% in 1995 to 30.56% in 2015, and the grey space expanded rapidly. Under the pressure of rapid urbanization, large areas of green land have become impervious surfaces, which is particularly evident in farmland and grassland area changes. Among the counted built-up areas in 2015, 125.92 km^2 of newly-added land came from farmland, and 563.41 km^2 of new land came from grassland. The percentage of farmland in the Wuhan metropolitan area dropped from 9.20% in 1995 to 6.0% in 2015. The rate of grassland in Wuhan metropolitan area decreased from 64.17% in 1995 to 45.36% in 2015. It is important to note that the forest area within the area is undergoing a process of fluctuating growth. The area had increased from 56.4 km^2 in 1995 to 64.3 km^2 in 2015. From this, we can see that during the process of urbanization in

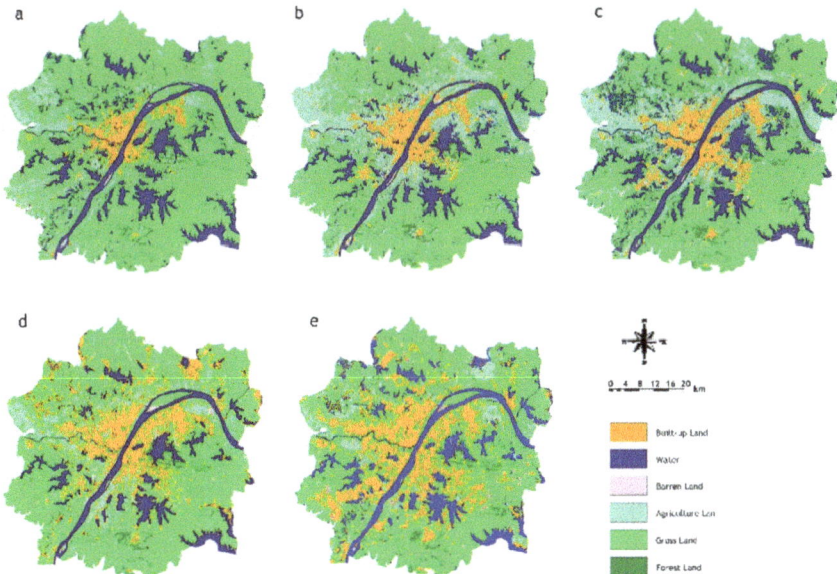

Fig. 4.5 Land use of the study area in **a** 1995; **b** 2000; **c** 2005; **d** 2010; **e** 2015

Table 4.3 Greenness areas in Wuhan metropolitan area from 1995 to 2015

Type	1995		2000		2005		2010		2015	
	Area (m²)	Rate (%)	Area (m²)	Rate (%)	Area (m²)	Rate (%)	Area (m²)	Rate (%)	Area (m²)	Rate (%)
Grassland	2063.8	64.1	1694.48	52.69	1651.01	51.34	1612.1105	50.13	1458.66	45.36
Forest	56.4	1.75	61.55	1.91	61.37	1.91	69.159164	2.15	64.3	2.00
Farmland	295.95	9.20	628.76	19.55	532.25	16.55	214.74746	6.68	192.82	6.00
Greenness	2416.15	75.1	2384.79	74.15	2244.63	69.80	1896.0172	58.96	1715.78	53.35

Table 4.4 Transition matrix of urban land use types in Wuhan metropolitan area from 1995 to 2015

Matrix Transfer Area	1995						
2015	Type	Water (km^2)	Forest (km^2)	Farmland (km^2)	Build-up land (km^2)	Grassland (km^2)	Wasteland (km^2)
	Water (km^2)	378.04	2.75	5.85	12.23	44.73	6.46
	Forest (km^2)	0.11	30.39	3.92	0.62	30.48	0.00
	Farmland (km^2)	37.97	0.39	39.88	12.28	104.84	1.13
	Build-up Land (km^2)	94.86	4.30	125.92	219.04	563.41	8.14
	Grassland (km^2)	37.78	16.96	184.02	56.58	1189.22	1.85
	Wasteland (km^2)	2.82	0.16	7.72	3.63	48.06	0.65

Wuhan, new construction area did not occupy the forest cover. Instead, 30.34 km^2 of grassland was converted into the forest.

From the perspective of the dynamics of different types of land cover changes (Fig. 4.6), the growth rate of the built-up area has a positive correlation with the decay rate of the green space, and the characteristics of changes in different time periods are different. From 1995 to 2000, grassland decreased at a rate of 21.8%, while farmland and built-up areas increased at a rate of 2.65% and 1.15%, respectively. From 2000 to 2005, grassland and farmland decreased by 2.63 and 0.91% respectively, and the built-up area increased by 0.13%. From 2005 to 2010, grassland and built-up areas increased at a rate of 2.57% and 2.84% respectively, and farmland decreased at a rate of 7.39%. While the grassland and farmland decreased at a rate of 16.17 and 0.57% from 2010 to 2015, the built-up area increased at a rate of 0.54%. Analysis shows that there is an internal transformation relationship between farmland and grassland. The rate of change of built-up areas and green areas is significantly related.

4.4.2 Circular Concentric Analysis

The circle analysis method analyzes the land use composition of each circle in Wuhan metropolitan area. Using seven circles in the Wuhan metropolitan area, we calculated the land use changes in each circle over 20 years. Further, the green space rate, the proportion of statistical green space accounting for the total area of Wuhan metropolitan area, is used to evaluate the change of green space in each circle. The

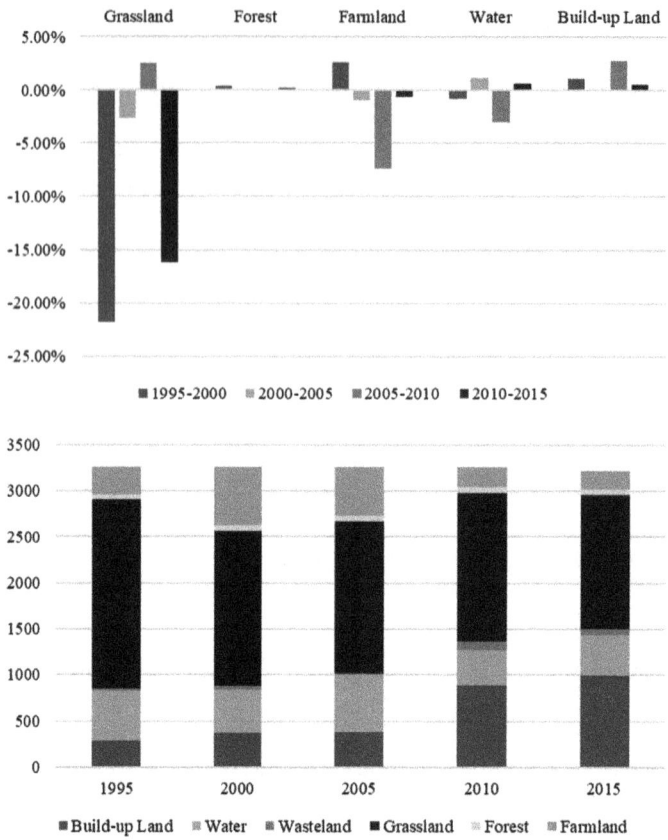

Fig. 4.6 Dynamics and composition of land use types in Wuhan metropolitan

results showed that the rate of green space in the urban areas of Wuhan increased from the center to the outside (Fig. 4.7). However, the change patterns of the various layers were different, and presented the following three rules: (1) The ratio of green space in the circle 1 was low but saw significant growth year over year. Circle 1 corresponds geographically to the Wuhan Second Ring Road. This area is the most mature area of urbanization in Wuhan. The grassland is mainly composed of parks, street green areas, etc.; (2) The ratio of green space in the circles 2–4 occurred in 1995–2010. The green areas have significantly decreased during the 15-year timeframe, but the rate of decline has slowed during 2010–2015. Circles 2–4 geographically correspond to the main urban area of Wuhan, which is the main area of urbanization in Wuhan. In 2005–2010, the amount of green space rapidly declined by about 15%?, but the declining momentum of green areas stopped after 2010; (3) Ring layers 5–7 slowly decreased year over year. Circles 5–7 correspond to areas within and outside the urban areas of Wuhan. Environmental quality of the region is best, but it shows a continuous decrease.

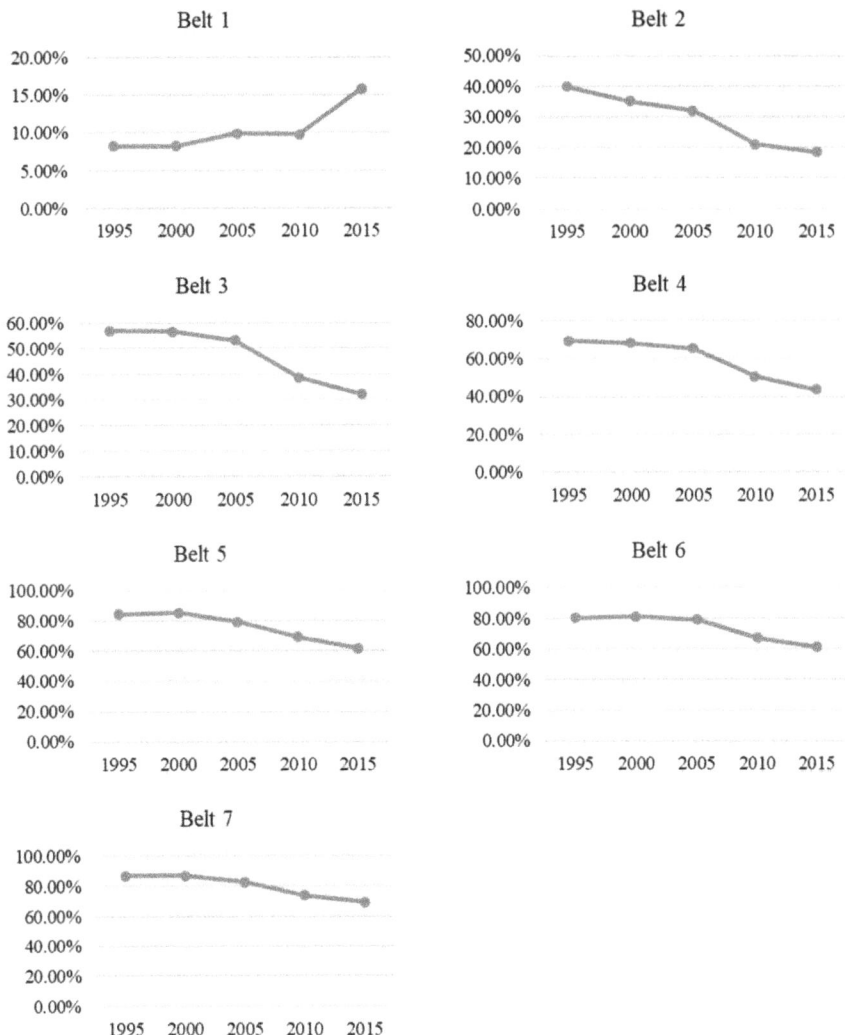

Fig. 4.7 Rate of change in green areas by ring layer in Wuhan Metropolitan area

4.4.3   Sector Directional Analysis

Sector fractals are used to analyze the changing patterns of green space in different directions. As shown in Fig. 4.8, the green area rate in most areas in the Wuhan metropolitan area in 1995 was above 70%. The highest rate of greenery in the south is 79.84%, and the lowest rate in the southwest is 66.30%. By 2015, the rate of greenery in the Wuhan metropolitan area is below 60%. The highest ratio of green space in the south is 58.94%, and the lowest ratio of green space is 42.50% in the southwest. Overall, the green space in eight directions of the Wuhan metropolitan area has

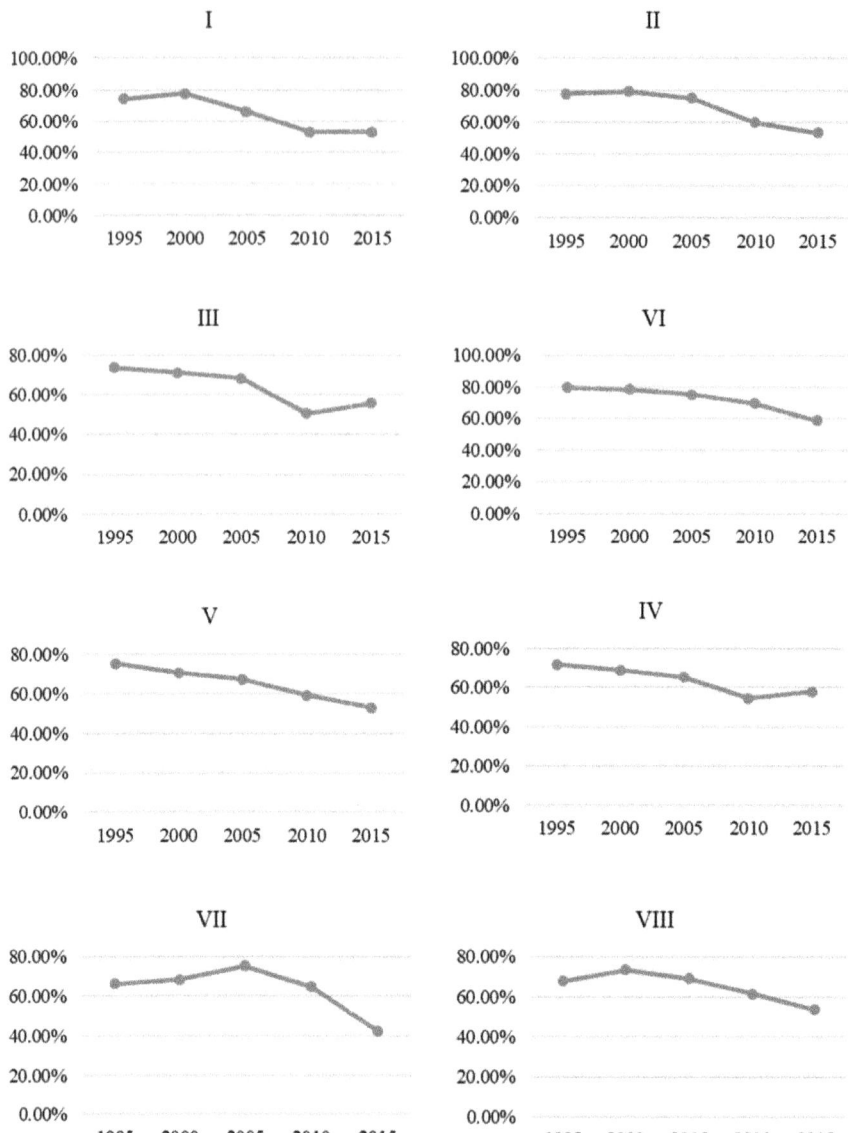

Fig. 4.8 Changes in greenness rates of sectors in Wuhan Metropolitan area

undergone a slow decay-rapid decrease-attenuation suppression process. However, this process is not continuous and each sector presents different characteristics.

We summarize the changes of sectors in the following four phases: (1) 1995–2000: The green space in the north of the Yangtze River grows slowly, and the green space in the south of the Yangtze River decreases slowly. During this timeframe, the green area rate of sectors I, II, VII, and VIII at this stage increased by about 2%, while the green area rate of sectors III, IV, V, and VI decreased by 1.27% to 4.77%. The green space has not yet significantly eroded; (2) 2000–2005: The green space in the entire region has significantly reduced; however, the green space in the southwest has slowly grown. Due to an increase in forest area, sector VII's greenfield rate increased by 7.03% at this stage, while other sectors significantly decreased at this stage; (3) 2005–2010: The region's green space was heavily affected. Due to the pressure of urbanization, the green space in the Wuhan metropolitan area decreased by 5.74–17.78%; (4) 2010–2015: The green space attenuation north of the Hanjiang River has stopped and has seen a gradually increase, while the green space south of the Hanjiang River has further decreased.

4.5 Discussion

4.5.1 The Evolution of Green Space and the Response of Policy Changes

From 1995 to 2015, the change of the green space in the Wuhan metropolitan area responded to the urbanization background and policies. The circle analysis shows the change in green space patterns from urban centers to urban fringe areas, reflecting the continuous expansion of the built-up areas of Wuhan metropolitan areas over the past 20 years. The development direction of sector fractals further illustrates that the southern areas of metropolitan areas and surrounding cities such as Huanggang and Xiantao are continuing to be built-up. The circle analysis shows that the number of greenbelts within the Second Ring Road of Wuhan Metropolitan Area has increased year over year which aligns with the demand for improved quality and quality of life in the area after becoming highly urbanized. The type of green space usually added is grassland (i.e. parks, community gardens, etc.), which coincides with citizen's daily life activities. The area outside the Second Ring Road, the main urban area of the Wuhan metropolitan area, experienced rapid decay before the decay rate began to decrease.

Around 2010, Wuhan successively issued relevant plans and regulations for ecological space protection (Table 4.5), which is in line with the desire to mitigate the loss of green land. The application of rigid planning methods in the main urban area usually yield good results. In areas outside the main urban area, green space is decreasing year over year, and this downward trend has continued. In the outer suburbs and marginal areas of the city, planning efforts are faced with much conflict due to more traditional protection and development thinking. Studies have shown

Table 4.5 Wuhan's plans and regulations on ecological spaces

Year	Title	Area	Management	Highlights
2001	Wuhan lake protection ordinance	Wuhan	Lakes	Management methods for lake bodies
20020	Wuhan city greening ordinance	Wuhan	Urban green space	Measures for the management of urban green space
2004	Wuhan forest resources management provisions	Wuhan	Forest	Management methods for forest land
2010	Wuhan East Lake scenic are master plan	East Lake	East Lake	The refinement of basic ecological control lines
2011	"1+6" space development strategy implementation plan for Wuhan urban development zone	Wuhan Metropolitan Area	Overall development	Consolidate the consensus of "great development" and "big ecology"
2012	Wuhan urban development zone 1:2000 basic ecological control line planning	Wuhan Metropolitan ecological area	Ecological space	Operable eco-line planning
2013	Wuhan greenway system construction plan	City of Wuhan	Urban greenway	Series ecological space, scenic tourism and comprehensive transportation
2013	Zhang Gongdi urban forest part construction plan	Zhang Gongdi	Zhang Gongdi	The refinement of basic ecological control lines
2014	Houguan Lake green wedge protection and development plan	Houguan Lake	Houguan Lake ecological space	The refinement of basic ecological control lines
2014	Yan Donghu—Yanxihu wetland park system planning	Yan Dong Hu-Yan West Lake	Yan Dong–Yan West Lake ecological space	The refinement of basic ecological control lines
2015	Wuhan city's global ecological framework protection plan	City of Wuhan	Ecological space	Determine the overall ecological framework structure

that urban planning actions face relatively slow execution and result in meagre practical effects in urban and rural fringe areas. It is difficult to create new environmental opportunities that simply rely on space policy tools (Shoard et al. 2002). The sectoral fractal analysis highlights the continuing trend of expansion of Wuhan metropolitan area to the. Since the northern part of Hankou is restricted by natural conditions, the built-up area needs to pay a higher economic cost to break through the highway. Therefore, the south of Wuhan metropolitan area bears a heavier burden to provide economic and structural development that depletes green space, despite the series of policies set forth to protect ecological space.

Different types of green spaces have experienced various development processes in the past two decades. In the process of urbanization, many open forest grasslands, water bodies, and farmland are transformed into dense, built-up urban areas. Sparse forest grasslands have continued to erod due to a lack of management and protection regulations. After the farmland undergoes rapid depletion, the downward trend is suppressed, thanks to the national policy of essential farmland protection. Since the consensus to protect forest and natural resources in Wuhan in recent years, these resources in the Wuhan metropolitan area have been efficiently and effectively protected. Forest areas are stable and slowly growing, while damaged mountains have since been repaired especially within the city. Additionally, after water resources in the urban area experienced the stage of "filling in the lake and building the city," Wuhan has successively protected lake resources such as Donghu, Yanxihu-Yandonghu and Nanhu Lakes, and the erosion of water bodies has stopped.

4.5.2 Targeted Strategies for Green Space Management

The overall protection of urban green spaces needs to reach social consensus. Currently, Wuhan metropolitan area still faces the dual pressures for social development and ecological protection. The question of whether the protection of green spaces will affect the future of housing and economic development should be further explored. Extensive and in-depth discussions to understand the direction and future goals of development will help lay the foundation for regional coordination and collaborative management A flexible and diversified management platform could be established to integrate multiple methods to solve the complex issues between the urban outskirts and periphery. The construction of ecological space should combine rigid and flexible principles. On one hand, efforts must explicitly propose to protect and guide the boundaries of urban growth. but they should also remain flexible When faced with uncertain socio-economic development needs in the future, space control should have the ability to adapt. In light of the changing characteristics of Wuhan's green space, it is recommended to adopt a combination of district and circle management methods. The main urban area is dominated by traditional planning methods, while in the outer suburbs, a flexible compensation and development mechanism is adopted.

4.6 Conclusion

In this study, circle analysis and sector analysis methods combined with interpretations of remote sensing images are used to describe the characteristics of green space changes in Wuhan metropolitan area from 1995 to 2015. The rate of change of different types of green space during the five periods of 1995–2000, 2000–2005, 2005–2010, and 2010–2015 was also used to calculate the change characteristics of the spatial composition of green space. The results of the study solved the first research problem. Rapid urbanization directly led to the loss of green space, especially for farmland and grasslands, while forests were less affected by urbanization. For the second research question, the study revealed differences in the variation of seven concentric circles and eight sectors. The rate of green areas in the core circle of Wuhan Metropolitan Area has continued to grow. The green area of the core area of Wuhan metropolitan area is low but continues to grow. The loss of the main urban area is severely controlled, and the marginal green space continues to be degraded and has no tendency to suppress. And the development trend is continuing. For the third research question, Wuhan Green Space has successively issued successful plans after undergoing rapid green space depletion and has seen good results in some regions and types of land. However, due to the lack of flexibility in control and failure to fully meet the actual development demands, planning failures have occurred in the urban fringe. This research shows that the combination of remote sensing interpretation and concentric and sector directional analysis plays an active role in the analysis of green space in Wuhan metropolitan area. The combined understanding of the changes in green space and time can help guide the governance of green space in metropolitan areas in the Midwest of China.

Acknowledgements We would like to give our special thanks to the following funding: (1) National Natural Science Foundation of China (Program Number: 51408248, 51478199, 51608213, 51778253); (2) China Scholarship Council (No. 201706165048); (3) The Recruitment Program of Global Experts (Youth Group) of China (Grant No. D1218006).

References

Attwell, K., Randrup, T. B., & Konijnendijk, C. C. (2000). Urban land resources and urban planting—Case studies from Denmark. *Landscape & Urban Planning, 52*(2), 145–163.

Baojun, Y., & Peng, C. (2015). The inheritance and transformation of urban planning under the new normal. *Urban Planning, 39*(11), 9–15.

Chao, C., Tao, J., & Yuanbin, Y. (2009) Application of supervised classification and visual modification in high-resolution remote sensing image. *Land Resources Information, 2009*(5), 37–40+48.

Cheng, J., & Masser, I. (2003). Urban growth pattern modeling: A case study of Wuhan City PR China. *Landscape and Urban Planning, 62*, 199–217.

Dickman, C. R. (1987). Habitat fragmentation and vertebrate species richness in an urban environment. *Journal of Applied Ecology, 24*(2), 337–351.

Fenghua, T. (2010). The formation mechanism and optimization of the spatial structure of urban ecological regions: An empirical study on the environmental green heart of Changsha, Zhuzhou, and Xiangtan urban agglomerations. China Urban Planning Society, Chongqing Municipal People's Government. Planning Innovation: 2010 China Urban Planning Conference Paper Set. China Urban Planning Society, Chongqing Municipal People's Government 2010:9.

Fernándezjuricic, E., & Jokimäki, J. (2001). A habitat island approach to conserving birds in urban landscapes: Case studies from southern and northern Europe. *Biodiversity & Conservation, 10*(12), 2023–2043.

Grove, J. M., Troy, A. R., Jpm, O., et al. (2006). Characterization of households and its implications for the vegetation of urban ecosystems. *Ecosystems, 9*(4), 578–597.

Hao, Z., Jixi, G., Gaodi, X., et al. (2012). Transpiration and temperature reduction functions of Beijing urban green space and its economic evaluation. *Acta Ecologica Sinica, 32*(24), 7698–7705.

Juan, S., Degao, Z., & Yi, Ma. (2014). Research on the characteristics and patterns of the development of near-space in megacities: Based on the discussion of Shanghai and Wuhan. *Urban Planning Journal, 06*, 68–76.

Kelly, K. M. (1999) *Urbanism, health and human biology in industrialized countries.* Cambridge University Press.

Kong, F., & Nakagoshi, N. (2006). Spatial-temporal gradient analysis of urban green spaces in Jinan, China. *Landscape and Urban Planning, 78*(3), 147–164.

Lang, Z. (2012). The construction of shanghai ecological network system based on organic evolution. *Chinese Landscape Architecture, 10*, 17–22.

Lanyun, Li., & Facheng, W. (2013). Ecological space planning management and guidance method in the main urban area of Chongqing. *The Planner, 29*(S2), 45–48.

Lin, Y. (2015). Controversies and Consensus in the Development of "Green Heart Strategy" of Randstad, The Netherlands in the 60 Years—Concurrently on the Enlightenment to Contemporary China. *International Urban Planning, 30*(06), 50–56.

Nagase, A., & Dunnett, N. (2012). Amount of water runoff from different vegetation types on extensive green roofs: Effects of plant species, diversity and plant structure. *Landscape and Urban Planning, 104*(3/4), 356–363.

Naveh, Z. (2000). What is holistic landscape ecology? A Conceptual Introduction. *Landscape and Urban Planning, 50*(1–3), 7–26.

Northam, R. M. (1975). *Urban Geography.* New York: J. Wiley Sons.

Onishi, A., Cao, X., & Ito T., et al. (2010). Evaluating the potential for urban heat-island mitigation by greening parking lots. *Urban Forestry & Urban Greening, 9*(4), 323–332.

Rees, W. E. (1997). Urban ecosystems: The human dimension. *Urban Ecosystems, 1*(1), 63–75.

Rui, L., & Dao-Lin, Z. (2010). Discussion on land use change information mining method based on transfer matrix. *Resources Science, 32*(8), 1544–1550.

Shoard, M., Edgelands, & Jenkins, J. (Ed.). *Remaking the landscape: The changing face of Britain* (pp. 117–146). London: Profile Books.

Sun, R. H., Chen, A. L., Chen, L. D., et al. (2012). Cooling effects of wetlands in the urban region: The case of Beijing. *Ecological Indicator, 20*, 57–64.

Tao, Y., Li, F., Wang, R. S., & Zhao, D. (2013). Research progress in the quantitative methods of urban green space patterns. *Acta Ecologica Sinica, 33*(8), 2330–2342.

Urs, P. K., Heather, G. H., Marty, D. M., et al. (2001). Change in ecosystem service values in the San Antonio area Texas. *Ecological Economics, 39*(3), 333–346.

Weber, T. (2004). Landscape ecological assessment of the chesapeake Bay Watershed. *Environmental Monitoring & Assessment, 94*(1–3), 39.

Xiaodong, L., Ming, Y., & Chaodong, H. Main features and thinking of the compilation of the master plan of the new edition of Beijing City. *Journal of Urban Planning, 2017*(5):41–49.

Zhang Biao, Xu., Jie, X. G., et al. (2016). Analysis on the pattern changes of urban green space in Beijing from 2000 to 2010[J]. *Ecological Science, 35*(6), 24–33.

Chapter 5
Relationship Among Fractional Vegetation Cover, Land Use and Urban Heat Island Using Landsat 8 in Taipei, Taiwan

Mu-En Chang, Zhi-Qing Zhao, and Hsiao-Tung Chang

5.1 Introduction

Extreme weather emerges recently and especially, the phenomenon of global warming gets focal attraction. The global atmospheric temperature has increased by 0.6 °C during recently one hundred years, caused by the greenhouse effect (Zhang et al., 2001). According to the file from World Meteorological Organization (WMO), cited from National Oceanic and Atmospheric Administration (NOAA), the average temperature of land and ocean is 0.76 °C during April, 2010 than that in twentieth century. It's pointed out from Environmental Change Research Project, conducted by Academia Sinica in Taiwan, that the average temperature has increased by 12 °C during the past one century for Taiwan and the amplitude of the temperature increase is 1.5 times than that worldwide, which reveals the severity of the abnormal weather in Taiwan Island.

The warming rate is higher in urban areas. The rate peaks at 4.2 °C every one hundred year during the past twenty five years (from 1969 to 1993) for Taipei City. 7 °C is for counterpart in Tokyo, Japan (Qiao, 1994). 61% of the population will live in urban areas by 2030 in the light of the statistics from the United Nations (U.N.) (UN, 1997). Urbanization is therefore the inevitable result of the development of social economy. The types of land use have been changed as urbanization develops. And the vegetable-dominant natural landscape is gradually replaced by artificial buildings, which further leads to the climate change including temperature increase and humidity decrease along with the green land reducing. This is just one of the elements for global warming.

The artificial buildings absorb large amount of solar heat during the day and release it during the night, making the city as a heat island. The phenomenon that

M.-E. Chang (✉) · Z.-Q. Zhao · H.-T. Chang
School of Architecture, Harbin Institute of Technology, Key Laboratory of Cold Region Urban and Rural Human Settlement Environment Science and Technology, Ministry of Industry and Information Technology, Harbin, China

© Springer Nature Switzerland AG 2021
W. Li et al. (eds.), *Human-Centered Urban Planning and Design in China: Volume I*, GeoJournal Library 129, https://doi.org/10.1007/978-3-030-83856-0_5

the temperature in urban area is higher than that in surrounding suburbs is called the effect of urban heat island. The most obvious in urban climate is that the temperature in urban areas is 2 °C than that in surrounding suburbs, which is also generally called effect of urban heat island (Oke, 1995; Taha, 1997).

As a special phenomenon, the effect of urban heat island is mostly affected by climate environment, urban heat environment and heat source, etc. (Saitoh & Hisada, 1991). In addition, the negative influence of the effect of urban heat island is numerous; including high temperature, aridity, sunlight reducing, and cloudiness increasing, slight increase of rainfall, reduce of average wind speed and air pollution, etc. (Lin et al., 1999).

Urbanization is the trend of social development, which causes huge impact on environmental temperature and global warming under the increasingly serious effect of urban heat island. The relationship between the effect of urban heat island and land use is revealed from the temperature and humidity files of the cities (Xiaoqing, 1998). The factors, which assess if the urban development is in accordance with the sustainable development of the environment, consist of resource utilization, concrete coverage, FVC, water circulation and air pollution (Lin, 2010). In other words, it's important to discuss the influence of the land use on green land and temperature. That is why the influence of land use on FVC and UHI is chose as the theme in this research.

It's consuming to traditionally measure the green land and collect the samples of vegetation both in money and manpower with possible huge difference, not suitable for large area of research. The remote sensing images can collect files within large area. In this project, the remote sensing technique is applied to extract light spectrum of the plants, which is established relationship with its vegetation coverage to further get FVC. The calculation of the intensity of urban heat island also deploys the satellite images to achieve the surface heat. And then the difference between the urban area with high temperature and suburbs without the effect of heat island is calculated to get the intensity of urban heat island (Lin, 1999). The division of land use is then analyzed in space using Geographical Information System (GIS) to further discuss the influence of land use on FVC and UHI and serve as serve as the reference for the urban design and the spread control of urban heat island in the future.

5.2 Materials and Methods

5.2.1 Study Area

Taipei, as the capital of Taiwan, is central to politics, economy, commerce and transportation, etc. This young metropolitan area includes mountain range, green range, watershed range and residence and commercial range in its abundant surface state. That is why Taipei is chose as the research base to test the surface temperature using TIRS image data (Fig. 5.1).

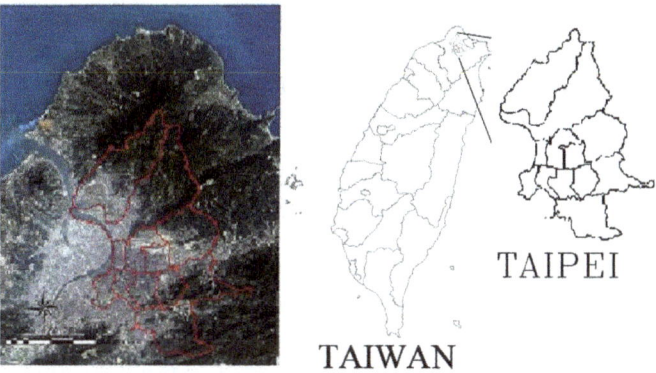

Fig. 5.1 Research scope in Taipei city

Taipei locates at the far north of Taiwan Island with the topography of basin, which speeds up in its development during the recent one hundred years, one of the representative cities during the period of colonial city. In 1895, Taiwan Island was ceded to Japan due to the defeat of China. And the Government House is set in Taipei City, becoming the political center of the Taipei City. Taipei City Plan was released in 1900 to vigorously transform Taipei City based on the modernized concept. Taipei City now has become the political, economic and trade center and most of the land is used as residence and commerce with high density and quality of office buildings. The temperature for Taipei City, like many other hot subtropical cities, is repeatedly soaring during recent years, breaking the record of the past years. It's much sultrier within the city because of the effect of urban heat island, greatly reducing the comfort of the city. Trees and green are insufficient due to the lack of space, resulting in a few degrees higher than suburbs.

5.2.2 Surface Temperature Calculated Using Satellite Images

Thermal infrared image of the remote sensing satellite has always been a major file to monitor the temperature change of the Earth's surface, offering an important source of information for the research of global climate change (Kustas & Anderson, 2009; Anderson et al., 2012; Georgescu et al., 2013). In the existing research of remote sensing of urban heat island, the main generation ways of Land Surface Temperature (LST) include grey level and brightness value of remote sensing images and remote sensing inversion, etc. there's a larger difference between grey level of surface features and brightness temperature and actual temperature of surface, not effective in supporting the quantitative study of the urban heat island. The LST-based inversion algorithm is proved as a surface temperature-acquired method with high precision to apply into the quantitative research of urban heat island. Landsat series of satellites has been playing an important role in the middle-scaled observation of surface using thermal infrared. Landsat8 or Landsat Data Continuity Mission (LDCM) mainly carries two loads: Operational Land Imager (OLI) and Thermal Infrared Sensor (TIRS), which has been improved significantly compared with previous Landsat series with the important mission: continuing the observations of earth thermal infrared from Landsat (Irons et al., 2012; Roy et al., 2014).

Landsat 8 TIRS thermal infrared sensor has two thermal infrared sensors: TIRS 10 and 11. OLI includes a full-color, eight multi-spectral, amounting to nine wave bands. Panchromatic resolution is 15 meters; multi-spectral resolution is 30 m; the width is 185 km and viewing angle is 15° (Yun, 2013; NASA, 2013; Zhang, 2013). Thermal infrared detection function is realized for TIRS to detect two thermal infrared radiations: 10.6–11.2 and 11.5–12.5 μm. The resolution is 100 m, viewing angle is 15°, ground width is 185 km, and quantitative bit is 12 bits. It can be used to monitor the utilization of land water resource. Two thermal infrared wave bands allows users to make atmospheric correction to thermal infrared data using split-window algorithms to further conduct better measurement to surface temperature (Jiang, 2013).

The temperature meteorological data of urban heat island originates from the calculation using the surface heat sensing from satellite. Land Surface Temperature is the important parameter to the research of regional thermal environment (Georgescu et al., 2013). The brightness temperature of surface, also known as surface radiation temperature refers to the temperature of radiation intensity observed by satellite sensor (Shan and Liangsong, 2012), which is the overall performance of surface temperature, surface emissivity, atmospheric transmittance and atmospheric temperature, etc. The calculation generally follows the Planck equation and calibration parameters. It's shown from many researches that the brightness temperature of surface contains more atmospheric information than surface temperature with the difference in magnitude, but the correlation is strong. The condition of spatial distribution also fit to between them to some extent, which effectively reflect the effect of urban heat island (Yongming et al., 2009). The specific steps of inversion for the temperature of surface brightness are shown as follows.

The research scope is the satellite image data in June 9th, 2015, which is extracted to analyze the surface temperature. And the resolution is 100 m from Landsat TIRS, which is transferred into surface temperature through the thermal image data corrected by emissivity (Nichol, 1994).

1. The estimation method of spectral radiance is conversed from DN of 10th and 11th of Landsat 8.

$$L(\lambda) = gain * DN + offset \qquad (5.1)$$

In Formula (5.1), it refers to the value of spectral radiance, the unit is $W \cdot m^{-2} \cdot sr^{-1} \cdot \mu m^{-1}$; DN is the value of remote sensing images; value is gain value; value is offset value. From the subsequent files of Landsat 8, it's investigated separately that the value *gain* is 3.342×10^{-4} and the value *offset* is 0.1.

2. The estimation method of Temperature kelvin is to Convert Radiance into degrees. Based on Planck equation, Kelvin converted separately into brightness temperature from the 10th and 11th wave band of Landsat 8 (Chander and Markham, 2003).

$$T_k = \frac{K_2}{In\left(\frac{K_1}{L_\lambda} + 1\right)} \tag{5.2}$$

In Formula (5.2), it's the radiance brightness; it's the spectral radiance; both are constant (Kelvin), From the subsequent files of Landsat 8, It's investigated that the value of the10th wave band K_1 is 774.89 (W · m^{-2} · sr^{-1} · μm^{-1}) and the value K_2 is 1321.08 (Kelvin), the value of the 11th wave band K_1 is 480.89 (W · m^{-2} · sr^{-1} · μm^{-1}), and the value K_2 is 1201.14 (Kelvin).

3. The estimation method of Land Surface Temperature is to convert from Temperature kelvin to Land Surface Temperature. It's only the absolute brightness temperature concluded from the above formula, which is under condition that any surface object is to assume the black body. However, the emissivity varies due to different surface objects in cities; therefore the surface temperature should be calculated based on the actual emissivity. The above formula is converted into:

$$T_s = \frac{T_k}{1 + (\lambda T_k/\rho) \ln \varepsilon_\lambda} \tag{5.3}$$

Formula (5.3) T_s is the surface temperature; T_k is the radiance brightness; λ is the length of radiation wave band; ρ is a constant; ε_λ is the surface emissivity. The lengths of radiation wave band are separately 10.3 and 11.5 μm for the 10th and 11th wave band in Landsat 8; Formula of ρ is $\rho = hc/\sigma$ (the value is 1.438×10^{-2} m · K); σ is the Boltzmann constant (the value is 1.38×10^{-23} J K^{-1}); h is Planck's constant (the value is 6.626×10^{-34} J · s); c is the velocity of light (the value is 2.998×10^8 m s^{-1}).

4. The estimation method of natural surface emissivity is closely related to NDVI. The surface pixel is mixed, which is constituted by surface composition with different portions. It can be divided into vegetation with large area or the built surface with bare soil, which can be directly represented by two types of surface emissivity. It's considered as the natural areas with abundant vegetation for the value of NDVI >0.157, the calculation formula is Van et al. (1993):

$$\varepsilon_\lambda = 1.0094 + 0.047\ln(NDVI) \tag{5.4}$$

It's considered as the built area with bare soil for the value of NDVI < 0.157, ε_λ ranges between 0.92 and 0.97. The calculation formula is:

$$\varepsilon_\lambda = 0.92P_v + 0.97(1-P) \tag{5.5}$$

Formula (5.5) is the FVC, also the constitution portion of the vegetation. The P_v formula is shown as follows.

$$P_v = [(NDVI - NDVI_s)/(NDVI_v - NDVI_s)]^2 \qquad (5.6)$$

Formula (5.6) is the minimized and maximized value of NDVI for vegetation and bare soil area (Owen et al., 1998).

5.2.3   Source and Methods of Files for Urban Heat Island

The degree of strength for urban heat island effect can be evaluated by a quantitative indicator, which is called the intensity of urban heat island, the difference between the maximum and minimum temperature of the central area and suburb (Lin, 1999). The methods are considerable to discuss the weather anomalies due to the effect of urban heat island. Willian Lowry put forward the model in 1977. In this model, three elements constitute Formula (5.7)

$$M = C + L + U \qquad (5.7)$$

C is the basic climate value of the region; L is the difference variable in its location; U is the changing items due to urbanization. These three elements are the variables to affect the city areas. M is the time series, which is the consolidated result of the statistics (Zheng, 1988).

In the Formula (5.8), U is the environmental difference between urban and rural areas to estimate and calculate the effect of urban heat island.

$$U = Mu - Mr \text{ (u is urban areas; r is rural areas)} \qquad (5.8)$$

Guandu Plain alluvial plain to the sea Tamsui Taipei city formed in the metropolitan Taipei area is the prohibition of the development, and therefore can be used as the outskirts of Taipei No heat island phenomenon survey point (Fig. 5.2).

Guandu Plain in suburbs is taken as the comparison area with no urban heat island. Guandu Plain is the alluvial plain formed by the sea buoy of Tamsui River in Taipei city, the prohibition development area in Taipei. It therefore can be the survey point with no heat island for outskirts of Taipei.

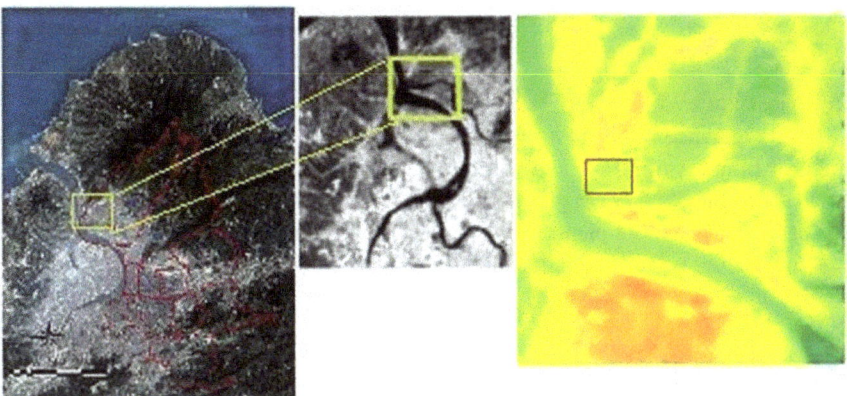

Fig. 5.2 Region selection without heat island effect in suburbs of the city

5.2.4   Source and Methods of FVC

The vegetation cover can affect the air quality and citizen's health (Wagrowski and Hites, 1997). The collection of vegetation samples traditionally is the sample estimation based on certain amount of sampling in research areas. The intensity of vegetation cover is estimated from the FVC of the samples, such as visual method and field survey. It's consuming in money and man-power due to the characteristics of the time and space with huge difference, which is unfavorable to the extraction of multitemporal vegetation within large scale. The application of remote sensing, however, can easily deal with the file collection within large scale along with the advantages of continuity of time and space. The files from remote sensing methods have already being dominant in the research of relationship between vegetation and climate. Spectral information of the vegetation is obtained using remote sensing technique to further establish the correlation with vegetation cover to ultimately generate FVC.

The formula of FVC, exponential equation, can be stated as follows using non-linear regression analysis (Agam et al., 2007)

$$FVC = 1 - \left(\frac{NDVI_{\max} - NDVI}{NDVI_{\max} - NDVI_{\min}} \right)^{\alpha} \tag{5.9}$$

In the formula, the exponent α is set at 0.625.

In the above formula, NDVI is the normalized index of different vegetation, the most commonly used indicators of green vegetation exploration, representing the degree of urban greening. NDVI is a good indicator to the coverage and growth state of macro-scope surface vegetation (Tucker, 1979). The concept of NDVI is defined as the ratio between the sum of near-infrared and red light and the difference between them (Rouse et al., 1973). The formula is shown as follows.

$$NDVI = (IR - R)/(IR + R) \qquad (5.10)$$

IR = near-infrared; R = red.

The reflectance value of the vegetation peaks at red and near-infrared and therefore the growth index of NDVI can be calculated to distinguish the plant and non-plant area. The value of NDVI ranges between $+1$ and -1. Greater index indicates denser of vegetation and more plants. The area usually is non-plant areas when the value is less than zero, such as cloud, water, roads and buildings.

5.3 Research and Analysis

5.3.1 Analysis of FVC for Taipei City

The files of NDVI (normalized difference vegetation index) serve as the comparison for the analyzed value of near-infrared and red band of satellite images and its conversion into FVC. The value is integrated into GIS, taking some blocks as unit to follow the view analysis of FVC (Fig. 5.3). The value of FVC takes some blocks as the average one in 2015. It's shown from the figure that the FVC for the northern and southern mountains and Guandu reach 50% with good condition, and FVC for other areas are in poor conditions. For the urban areas, in particular, FVC is all lower than 20% and many areas are even lower than 10%. And only a small part of areas within the cities is higher than 30% in FVC. The areas with FVC lower than 20% mostly is the city center dominated by artificial items. A few green lands with FVC lower than 30% scatter randomly, failing to reach the temperature reducing in series as the green net.

5.3.2 Analysis of Urban Heat Island in Taipei City

Urban heat island is calculated and analyzed for Taipei city using the satellite Landsat 2015 (Fig. 5.4). From the figure, it's displayed that the central areas is the most serious areas, all higher than 2 °C and it's even higher than 8 °C. The eastern and southern direction is in special serious condition. This shows that Xinyi District in east part and Nangang science and Technology Park are both well-developed. The business center of Taipei City moves to the east. The central commercial areas and the administrative buildings of Taipei City Government all are moved to the east, which brings the complex political, commercial, trade and transportation behaviors. The speeding development also aggravates the urban heat island for the east of Taipei City. Cool Island phenomenon emerges due to the mountainous terrain for the northern part; some areas are found Cool Island in western part and the water flowing is the possible reason to mitigate urban heat island.

Fig. 5.3 Distribution of
FVC for Taipei city

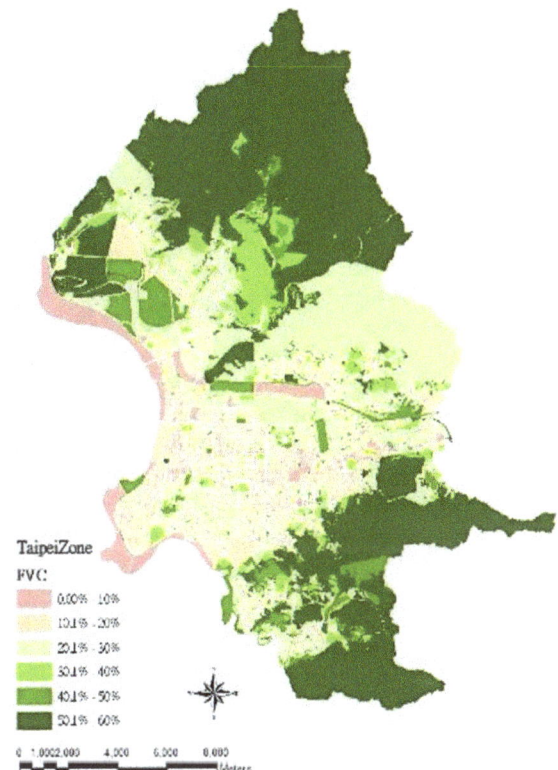

5.3.3   FVC of Taipei City and Analysis of Urban Heat Island

In this study, some blocks are considered as unit and GIS as platform to match FVC
with UHI to follow further analysis. The number of files is 13,854. The land use is
classified into 53 types to conduct statistics and average. FVC and UHI are analyzed
from each kind of land use, following by the average of two items.

The fitting test of multiple regression models is conducted using Excel software
to simulate the environment evolution curve and the trend line is applied to get to
know its regression equation (Fig. 5.5). The model selection is chose following the
simplest principles if several models pass the examination, such as simple variation
and easy subsequent analysis. The transition model of each environment quality is
listed in the following table through the observation of R^2 and the value of P (Table
5.1).

From the above figure, it's shown that the relationship between FVC and UHI
displays the inverted U curve to the right; binomial regression is the most reliable
from the table. FVC has a significant negative correlation with UHI for Taipei City.

Fig. 5.4 Distribution of UHI
for Taipei City

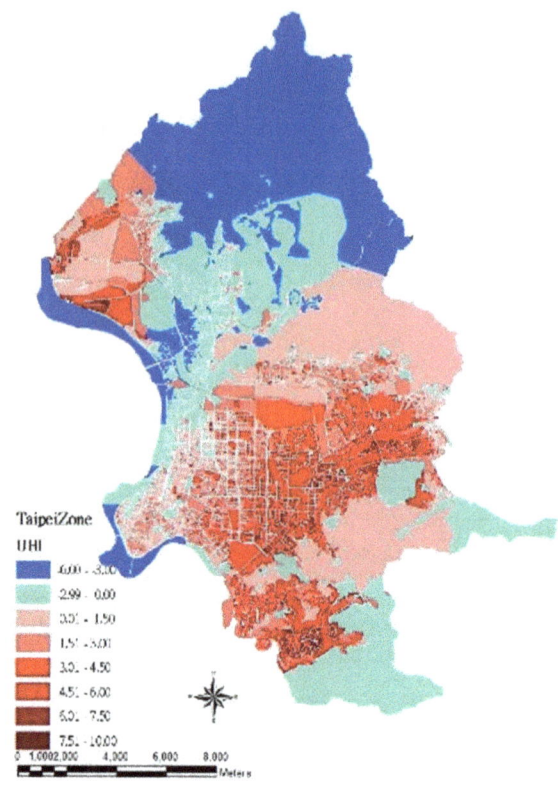

TaipeiZone
UHI
- -6.00 - -3.00
- -2.99 - 0.00
- 0.01 - 1.50
- 1.51 - 3.01
- 3.0. - 4.50
- 4.5. - 6.00
- 6.0. - 7.50
- 7.5. - 10.00

0 1.0002,000   4,000   6,000   8,000
                                        Meters

Fig. 5.5 Regressions of UHI
and FVC

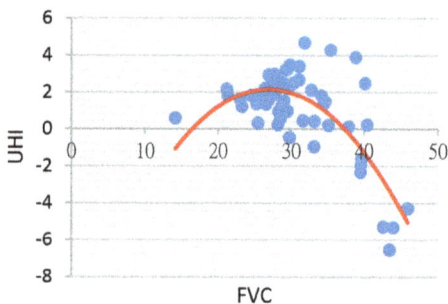

5.3.4   Analyses Between Land-Use-Based FVC and UHI for Taipei City

In this research, FVC of land use and change of UHI are also analyzed (Table 5.2). The average of all FVC is 28.25% and the average UHI of land use is approximately 1.87 °C, which shows the insufficient FVC has impact on UHI for Taipei City. The city center is higher in temperature with lower FVC and obvious UHI; the temperature

Table 5.1 Collection of regression data

	Coefficient	Standard deviation	t statistics	P-value
X^2	−0.019	0.004	−5.362	1.83E-06
X	1.064	0.239	4.454	4.38E-05
Intercept	−12.142	3.795	−3.200	0.002
F(S)	33.214			
R^2	0.539			
Standard deviation	1.555			

for northern and southern mountain area is lower with higher FVC and UHI not obvious.

From the perspective of land use, UHI peaks at 3.84 °C for special use area and 2.99 °C for industrial land. The average value of FVC for this two are 29.84 and 29.95%, not reaching to 30%. The special use are refers to the mixture of industry and commerce. It's concluded that the industrial land use easily leads to the emergence of UHI. The lowest average intensity of UHI are −5.31 °C, national park and −0.45 °C, scenic areas, both of them are Cool Island areas. The average values of FVC for this two are the highest, 43.97 and 31.61%. The national park land accounts for 19.71% of Taipei land use, which is the most land use to control the phenomenon of UHI. However, the phenomenon of UHI exist in almost all and use, except for only a small portion without UHI, such as coal and gas land, national park and scenic land use.

Residential area accounts for the most in architectural land use, 16.02%. The requirement for life quality and green land has been improved through the green land and the FVC has reached to 34.73%. The control of residential area to UHI has reduced to 0.41 °C, a good phenomenon and trend. The schools and government office land use has serious UHI with large area and for former, 2.12 °C and 32.75%; for latter 1.53 °C and 29.03%. The obvious improvement of FVC hardly reflects the control to UHI. For the commercial and market area with commercial behavior, the value are 2.00 and 2.49 °C in respective, all exceeding the average value of UHI of Taipei City. The value of FVC is 27.69 and 30.27%. The big problem is the control to UHI, calling for prioritized solution.

The highway is the lowest in UHI for transportation land use with highest FVC, 0.20 °C and 35.09%. The UHI for railway and aircraft land use all exceeds the average of Taipei City and the FVC all lowers the average, for railway land use, 2.53 °C and 27.41%, for aircraft land use, 2.94 °C and 26.93%. These two transportation behaviors all easily cause the serious phenomenon of UHI.

The phenomenon of UHI exists in all recreational land use except for the national park and scenic areas. The conservation area also has UHI, 0.12 °C even it accounts for 27.81% of the city area. It requires more cautious to the plan of conservation area in the future. For the cities, the UHI is serious for the area offering the green land environment, including squares, amusement area, entertainment area and park. UHI for square is 1.59 °C, amusement area, 1.39 °C and FVC is 34.46%; entertainment

Table 5.2 Comparison of UHI and FVC for city land use

City land use	Land use	UHI (°C)	FVC (%)	Proportion (%)
Architectural land use	Residential area	0.41	34.73	16.02
	Commercial area	2.00	27.69	3.24
	Storage facility	0.59	14.15	0.00
	Government office land use	1.53	29.03	3.52
	School	2.12	32.75	4.56
	Cemetery	0.23	40.43	1.25
	Cola and gas land use	−0.05	33.11	0.01
	Waste water treatment	1.22	23.29	0.63
	Market	2.49	30.27	0.22
	Historical sites	1.00	29.47	0.13
	Gas station	1.38	25.30	0.04
	Transformer station	1.79	25.86	0.05
Transportation land use	Railway land use	2.53	27.41	0.08
	Aircraft land use	2.94	26.93	0.99
	Parking	1.62	23.02	0.11
	Highway	0.20	35.09	1.30
	Others	1.82	24.52	0.63
Recreational land use	Square	1.59	26.95	0.11
	Amusement area	1.39	34.46	1.04
	National park	−5.31	43.97	19.71
	Entertainment area	2.05	27.84	0.39
	Scenic area	−0.45	31.78	0.75
	Conservation area	0.12	37.89	27.81
	Park	1.36	26.55	5.95
Industrial land use	Industrial area	2.99	29.95	1.9
Agricultural land use	Agricultural area	0.45	31.61	1.91
Special land use	Specific area	3.48	29.84	0.01
Water conservancy land use	Dike land	0.25	28.20	0.61
	Water line area	0.33	25.46	6.91
Others	Others	2.67	31.07	0.14

area, 2.05 °C, FVC, <27.84%; Park. 1.36 °C, 26.55%, indicating serious insufficient of green land. This kind of land use is lower for FVC and higher for UHI in all land use.

For water line and dike area, UHI is lower due to the influence of river, 0.33 and 0.25 °C in respective. Both FVC all lowers than 30, 25.46 and 28.20%. Temperature is hardly reduced to achieve Cool Island effect due to the phenomenon of urban heat island.

5.4 Discussion

1. In this research, the image data of Landsat 2015 is applied to discuss the FVC and UHI. Landsat satellite has been applied widely and it's proved highly creditable in the research of green land and UHI through the comparison between its file and the actual data.
2. It's concluded from the result that FVC has a negative relationship with UHI for Taipei City. The average value of FVC is 28.25% and the average intensity of UHI is about 1.87 °C for each land use.
3. The temperature of city center is higher with low FVC, showing obvious effect of urban heat island for Taipei City; Xinyi District in east part and Nangang science and technology park are both well-developed to speed up the development of Eastern Taipei City with more serious phenomenon of UHI. Cool Island phenomenon emerges due to the mountainous terrain for the northern part; some areas are found Cool Island in western part, which may be due to the mountain and water-oriented plan.

5.5 Conclusions

We are able to understand the influence of land use on the temperature and various types of land use deeply affect the changes of city temperature. It's concluded from the above research that FVC is insufficient, which is related to the temperature increase and higher UHI.

The mixture land use of industry and commerce, the special land use and industrial land use is the most serious in UHI with the FVC <30%, which shows that the plan and position for this kind of area should be more careful with the priority of temperature cooling and greening.

It's found that the commercial area and market in architectural land use is relatively higher in UHI than others. UHI is improved due to the attention to the life quality for residential area. It's worthy to mention that the government office land use and schools, which is under the jurisdiction of the government, are in server condition in UHI although FVC is higher. The improvement of FVC probably lies in the promotion for the green architecture guidelines and rules from Taiwan government and the regulation that the new buildings should follow the levels of the guidelines and rules. The green architecture therefore effectively improves the FVC of governmental buildings, but there are still considerable old government offices and units without regeneration. The average UHI is therefore relatively serious. It's estimated that the renovated buildings definite serve as the modes to reduce UHI in schools and government land use with the cooperation with sustainable implementation regulated by the rules of green architecture.

For transportation land use, only the FVC of highway is higher with lower UHI. Others are all lower in FVC and higher in UHI. The lowest FVC is parking lot and UHI peaks for aircraft land use. It reflects that the greening of transportation for

Taipei City is lower, which is one of the elements to affect UHI. The focus for the plan in the future is the greening of the road within the city.

The FVC for national park and scenic area land use all exceeds 35% without UHI. For agricultural and conservation area, natural green land is more with lower UHI and higher FVC, which is beneficial to control the city temperature. However, the FVC for park green land, entertainment area and squares all is lower than 30% with higher UHI. This area should be the open space for the citizen leisure, which conversely leads to the emergence of UHI. The main cause for constant UHI is therefore found out. The parks in Taipei City are mostly amusement or comprehensive type, where there's too many artificial and entertainment facilities. The widely used concrete and bricks for pavement lead to the lack of green land. Many neighborhood parks scatter in the corners, hardly reaching large area or the effective green corridors in the form of series. The amusement land use hardly controls the UHI within the cities.

The above result is the condition of UHI and FVC of Taipei City under the former urban plan, which can offer a result assessment for the former urban plan and governmental policies. The improvement focus and suggestions are also put forward for the urban regeneration and urban design. The control of city greening on UHI is explored for the governmental departments and academic filed to establish a healthy city, which mainly aims to mitigate the hurt of environment from the city development and the strategies is in planned to deal with the environmental problems. Taipei is condensed in population with abundant commercial behavior and the city development is the main factor to reduce the plants. It's difficulty to add large-scale green open space for the center of Taipei City with highly value of land. Concept of green corridor is the one solution. The neighborhood parks are connected in series though the road greening to constitute the condensed green corridor in the city center. The artificial items become greening outside. And the riverbank is also connected with communities in series. Besides, the mountains are conserved in sustainable ways. This action can be helpful to achieve green Taipei in the future, no longer a dream.

References

Agam, N., Kustas, W. P., & Anderson, M. C., et al. (2007). A Vegetation index based technique for spatial sharpening of thermal imagery. *Remote Sensing of Environment, 107*, 545–558.

Anderson, M. C., Allen, R. G., Morse, A., & Kustas, W. P. (2012) Use of Landsat thermal imagery in monitoring evapotranspiration and managing water resources. *Remote Sensing of Environment, 122*(SI), 50–65.

Chander, G., & Markham, B. (2003). Revised Landsat-5 TM radiometric calibration procedures and postcalibration dynamic ranges. *Transactions on Geoscience and Remote Sensing (IEEE), 41*(11), 26742677.

Gaozhen, J., Bing, H., & Yingbo, G., et al. (2013). Review of 40-year earth observation with Landsat series and prospects of LDCM. *Journal of Remote Sensing, 17*(5), 1034–1047.

Georgescu, M., Moustaoui, M., Mahalov, A., & Dudhia, J. (2013). Summer-time climate impacts of projected megapolitan expansion in Arizona. *Nature Climate Change, 3*(1), 37–41.

Guo, X. (1998). *The analysis of urban heat island effects of the Taichung City*. Master thesis. Department of Architecture of National Cheng Kung University, Tainan, Taiwan.

Irons, J. R., Dwyer, J. L., & Barsi, J. A. (2012). The next Landsat satellite: The Landsat data continuity mission. *Remote Sensing Environment, 122*, 11–21.

Jiang, S., & Zha, L. (2012). The quantitative relationship between land surface temperature and land cover types based on remotely sensed data in Hefei city. *Journal of Anhui Normal University (Natural Science), 35*(3), 252–257.

Kustas, W., & Anderson, M. (2009). Advances in thermal infrared remote sensing for land surface modeling. *Agricultural and Forest Meteorology, 149*, 2071–2081.

NASA.LDCM Launch. [EB/OL]. (2013). http://www.nasa.gov/mission_pages/landsat/launch/index.html.2013.

Lin, X. (1999). *Urban and rural ecology*. Taiwan: Jane's Bookstore.

Lin, J. (2010). The influence and environmental meaning of urban heat island effect. *National University of Tainan (Journal of Ecology and Environmental Sciences), 1*(3), 1–15.

Lin, X, Li, K., Chen, G., Lin, L., Guo, X., & Chen, Z. (199). Experimental analyses of urban heat island effects of the Four Metropolitan Cities in Taiwan (I)—The omparison of the heat island intensities between Taiwan and the world cities. *Journal of Architecture, 31*, 51–73.

Nichol, J. E. (1994). A GIS-based approach to microclimate monitoring in Singapore's high-rise housing estates. *Photogrammetric Engineering and Remote Sensing, 60*, 1225–1232.

Oke, T. R. (1995). The heat island of the urban boundary layer: characteristics, causes and effects. In: J. E. Cermak (Ed.), *Wind climate in cites* (pp. 81–107). Netherlands: luwer Academic Publishers.

Owen, T. W., Carlson, T. N., & Gillies, R. R. (1998). Assessment of satellite remotely sensed land cover parameters in quantitatively describing the climatic effect of urbanization. *International Journal of Remote sensing, 19*, 1663–1681.

Qiao, O. H. (1994). *The initial investigation and research of urban warming for Taipei City*. Executive Environmental Protection Agency, Taipei.

Rouse, J. W., Haas, R. H., Shell, J. A., & Deering, D. W. (1973). Monitoring vegetation systems in the Great Plains with ERTS-1. In *Third Earth Resources Technology Satellite Symposium* (309–317).

Roy, D. P., Wulder, M. A., Loveland, T. R., Woodcock, C. E., Allene, R. G., Anderson, M., & Helderg, D. (2014). Landsat-8: Science and product vision for terrestrial global change research. *Remote Sensing of Environment, 145*, 154–172.

Saitoh, T. S., & Hisada, T. (1991). Reduction of air pollution by changing the pollutant emission from the vehicles. In Proceedings of 26th IECEC'91 (Vol. 6, pp. 126–131).

Taha, H. (1997). Urban climates and heat islands: albedo, evapotranspiration, and anthropogenic heat. *Energy and Buildings, 25*, 99–103.

Tucker, L. I. (1979). Red and photographic infrared linear combinations for monitoring vegetation. *Remote Sensing of Environment, 8*, 127–150.

UN (United Nations). Urban agglomerations. (1996). *UN Population Division*. New York: United Nations.

Van De Griend, A., & Owe, M. (1993). On the relationship between thermal emissivity and the normalize difference vegetation index for nature surfaces. *International Journal of Remote Sensing, 14*(6), 11191131.

Wagrowski, D. M., & Hites, R. A. (1997). Polycyclic aromatic hydrocarbon accumulation in urban, suburban and rural vegetation. *Environmental Science and Technology, 31*, 279–282.

Ying, Y. (2013). The United States earth observation satellite Landsat-8. *Satellite Applications, 2*, 76.

Yujun, Z. (2013). Landsat 8 profile. *Remote Sensing for Land-Resources, 25*(1), 176–177.

Xu, Y., Tan, Z., & Zhu, Y. (2009). Spatial and temporal analysis of urban heat island in Suzhou city by remote sensing. *Scientia Geogreaphica Sinica, 29*(4), 529–534.

Zhang, R., Zeng, J., & Gu. X. (2001). The initial application of A four-dimensional variational assimilation prototype system to improve the track forecast of typhoon. In Papers Compilation of 2001 Conference on Weather Analysis and Forecasting (pp. 35–44).

Zheng, S. (1988) *Urban climatology*. Taiwan: Xu's Foundation.

Chapter 6
Ecological Perspective of the Evolution of Urban Spatial Form and Construction: Case Study of Hefei City

Dazhi Gu and Huifen Huang

6.1 Introduction

Under rapid urbanization, the urban construction speed in China consistently accelerates with increasing population and expanding urban scale. The continuous development of urban construction has resulted in the decrease in quality of the ecological environmental and target economic benefits. Thus, the ecological environment has become a critical constraint that limits the development of urban space and economy. Accordingly, the establishment of an ecologically healthy urban space and realization of urban sustainable development are among the important tasks of humanity. The essence of urban spatial development is the continuous invasion of natural ecological space through artificial constructions and the spatial interpenetration of natural elements and the city. The current ecological environmental problems that all large cities face originate from the lack of attention to natural ecological system. Many experts have proposed related suggestions, strategies, and theories, such as Howard's Garden City, Wright's Broadacre City, and Saarinen's Organic Decentralization, to solve these problems. These ideas are planning theories that comprehensively consider the natural ecological space and urban construction and attempt to strengthen the connection between these two concepts. The common point of these theories is that they emphasize the importance of ecological concept and the buffer and fusion of the ecological environment toward the urban artificial material environment. However, these theories have limitations because they are generally conceptual understanding that lack scientific direction and have yet to be implemented in actual situations. In recent years, the development of ecology and its intercross with other

D. Gu (✉) · H. Huang
School of Architecture and Arts, Hefei University of Technology, Shushan District, No. 485
Danxia Road, Hefei 230009, Anhui, China
e-mail: 67134971@qq.com

H. Huang
e-mail: 1072933453@qq.com

© Springer Nature Switzerland AG 2021
W. Li et al. (eds.), *Human-Centered Urban Planning and Design in China: Volume I*,
GeoJournal Library 129, https://doi.org/10.1007/978-3-030-83856-0_6

disciplines has enabled the former to serve as the bridge that links various disciplines. Moreover, studies on the relationship between the natural ecological elements and urban spatial development, such as the establishment of urban ecology and proposed "Shan-shui city" concept, are increasing. The natural ecological elements should be integrated into the fields of urban layout and urban designing, thereby providing a theoretical and technological guide for the study on urban spatial development from the ecological perspective (Aifeng and Yongchao, 2015).

This study uses urban planning over the years as basis to study the Hefei urban area using the induction–deduction method and literature review through the natural ecological perspective to determine its developing principle. This research combines the present construction situation of Hefei in proposing the corresponding suggestions and strategies from the macro, meso, and micro levels for the problems existing in Hefei's urban spatial construction from the ecological perspective.

6.2 Evolution of Urban Spatial Morphology in Hefei from the Ecological Perspective

6.2.1 General Situation of Hefei City

Hefei City is the political, economic, cultural, educational, commercial, transportation, and information center of Anhui Province. This city is located at the center of Anhui beside Chaohu Lake. Hefei lies in the central part of a hilly area in East China between Yangtze River and Huaihe River and the west end of the Pan-Yangtze River Delta Region. This city strategically links China's east and west and south and north regions. The total area of Hefei is 7029.48 km^2 (including 72.93 km^2 of the water surface area of Chaohu Lake), the population is 4.8674 million, and the urbanization rate is 62.4%. This city governs the four districts of Yaohai, Luyang, Shushan, and Baohe; and the three counties of Feidong, Feixi, and Changfeng (Hefei Urban Planning Bureau 1958, 1979, 1995, 2006, 2011).

From a macro location, Hefei is located in Central China and lies near the Yangtze River and is an important member of the Yangtze River Economic Zone.

Moreover, Hefei's location between the Yangtze River and Huaihe River has enabled this city to connect the southern and northern regions of China. In 2016, Hefei was designated as a sub-center among the city groups in the Yangtze River Delta Region. In the future, the continuous development and improvement of transportation facilities and the regions emergence will enable Hefei to become a crucial joint of the arterial railway system that links Eastern and Western China, the arterial highway system that links the country's southern and northern regions, and the arterial inland waterway system. Hefei also serves as the geometrical center of Hefei Province and the center of transportation system within the province. Given the developing opportunity brought about by its advantageous location, Hefei City also functions as a leader in the economic development in the province (Fig. 6.1).

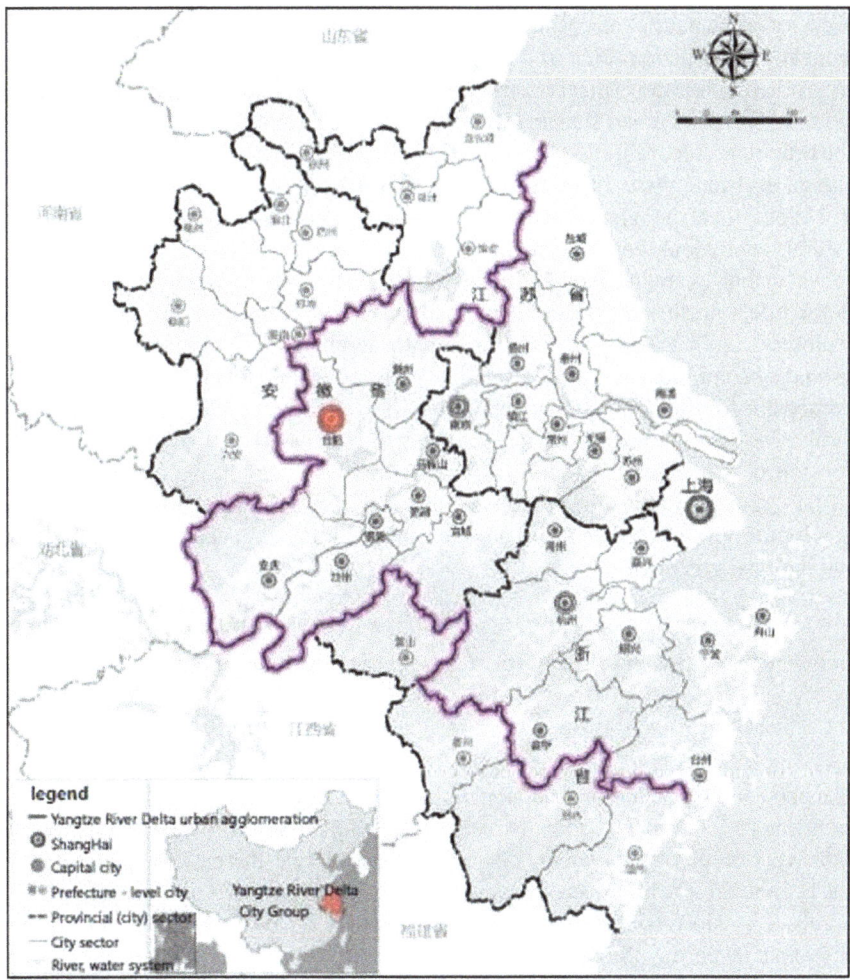

Fig. 6.1 Yangtze River Delta city group scope map (Painted by author)

6.2.2 Influence of Natural Ecological Elements on the Urban Spatial Morphology and Layout of Hefei

Hefei's Advantage in Natural Resources

The natural resources of a particular geographical location, topography, climate, and hydrology generally form the basis of the appearance, development, and construction of a city. These factors connect closely with a city's future development potential, spatial development direction, construction speed, land development intensity, overall layout of urban space, and interior structure. Moreover, the critical bearing capacity of the natural environment is a key reference foundation for urban development and construction. That is, the development and construction of a city should be controlled within the critical bearing capacity of the natural environment to ensure the orderly, healthy, and sustainable development and avoid blind development and construction. Accordingly, this type of development and construction may cause natural disasters, waste of resources, and deterioration of environmental quality.

Hefei's ideal geographical location has provided this city with advantageous natural, ecological, and spatial resources, such as forests, vegetation, and bodies of water. For example, Dongpu Reservoir is contiguous to Dafangying Reservoir in the city's northwest, Dashu Mountain and Zipeng Mountain are located in the southwest, and Chaohu Lake is located in the southeast. The mountains and waters around Chaohu Lake has formed an urban green spatial system that formed the basic framework for the urban green-land system in Hefei, thereby linking the urban and rural areas. Moreover, the interior waterway in Hefei's urban area of Hefei is well-developed, while the "Metropolis–Binhu District" is the intensive representation of the lakefront characteristic of the city Hefei. These conditions provided a good condition for the development of an urban lakefront greenbelt, greenbelt in Binhu District, and regional greenbelt.

Deficiencies of the Current Natural Resources Situation in Hefei

(1)   Scarce greenbelt for living

The passage of time has prompted many cities to considerably focus on the environment and Hefei is no exception. In recent years, Hefei has reinforced its construction endeavors on city parks, greenbelts, and squares, while the environmental quality has been relatively improved. However, the construction speed of greening in Hefei is relatively slow compared with those in other second-tier cities with the same economic level (e.g., Suzhou, Nanning, and Kunming). The development of greenbelts within Hefei is unbalanced, lacks a uniform planning of the entire city greening, the comprehensive benefit has not made its full play, and the green rate per capita is relatively lower than that in other cities.

(2)   Serious pollution of water resource

Hefei is abundant in water resources. The city's natural lake (i.e., Chaohu Lake) and two artificial reservoirs (i.e., Dongpu Reservoir and Dafangying Reservoir) provide domestic, industrial, and irrigation water for the city. Nanfeihe River and Shiwulihe River run through the urban area and flow into Chaohu Lake. The discarding of household garbage and emission of industrial sewage decreased the water quality in Hefei and caused the serious pollution of water resources.

(3)   Reinforcing the construction of forest ecological system

"Trees around the city, greenbelt through the city, interdependent forest and water, harmonious forest and city" are among the characteristics that shaped the urban spatial morphology of Hefei. The construction of city forest ecological system cannot be separated from the establishment of low carbon city. Although Hefei has abundant forest resources, the consciousness related to its protection and sustainable development has not been promoted. Therefore, Hefei should continuously innovate and develop relative concepts, strengthen the integrated rural–urban greening planning; remain with the "trinity" of the city, suburb, and country, "trinity" of networks of forest, water, and road, and "trinity" of the construction of the ecological, economic, and landscape forest; enrich the form and content of forest planting and greening; improve the ecological quality; and make full play of the ecological, social, and economic benefits of forests (Liang et al., 2015).

Limitation of Ecological Elements Toward the Urban Spatial Form Evolution in Hefei

Many factors influence the development and construction of a city. Similar to politics, economy, and culture, the natural ecological element is one of these factors. The close connection between Hefei's urban spatial development and natural ecological elements has been determined through this city's urban planning organization over the years. Through a panoramic view of the entire urban space, the main elements that influence urban spatial morphology in Hefei include the Dashu Mountain, Zipeng Mountain, Dongpu Reservoir, Dafangying Reservoir, Chaohu Lake that locate in the city domain and the greenbelt vegetation space within certain area around the mountains and waters; numerous parks; and public greenbelts along Huancheng Park within the urban area. These natural ecological elements of mountains, waters, and greening spaces limit the blind chaotic expansion of the city through their influence on the urban spatial morphology in Hefei.

6.2.3   Evolution of Urban Spatial Form in Hefei from the Ecological Perspective

Hefei City has occupied a specific surface area since its establishment and formed certain outline and form for land-use under the restriction and influence of the types of natural and artificial factors. The urban spatial form can be divided into the following main types based on its planned shape of the main body of the urban built-up area: focal, linear, radial, cluster, conurbation, and scattered forms (Qibiao, 2015) (Table 6.1).

Table 6.1 Types of urban spatial form (Painted by author)

Form symbol	Form name	Form symbol	Form name
⊕	Focal form	⁂	Cluster form
⬡≋≋	Linear form	⁂	Conurbation form
⬡	Radial form	o o / o o o / o o	Scattered form

Since the designation of Hefei as the capital of Anhui Province in 1950, the city has achieved a continuous acceleration of urban development and expansion of urban scale. After the Liberation, six rounds of urban planning have been conducted under the direction of the Draft on the Planning of Hefei Streets issued in 1953. The current study divides Hefei into two periods based on the development of the strategic layout of Hefei, assesses the six rounds of urban planning from the ecological perspective, and concludes the types and characteristics of its urban spatial form (Qianqian et al., 2011).

Around-City Period

Urban construction in Hefei was previously slow. Moreover, the urban scale of this city was small before the Liberation with a population of 50,000. Meanwhile, the urban built-up area of below 5 km^2 was limited in the present old city area. In the early years after the Liberation, the urban layout was in a three-leaf-fan form because of the limited natural factors. These limiting factors are mainly from the Nanfeihe River, Dongpu Reservoir, and Dafangying Reservoir. The terrain in southeast Hefei is substantially low and located at the lower reaches of the Nanfeihe River. Hence, the geology in this area is not qualified for urban construction. Given the damage that flooding can cause, certain areas around the Dongpu Reservoir and Dafangying Reservoir designated as construction restricted area. In the future, Hefei City will generally develop toward the east, north, and southwest, thereby strengthening the foundation for the future development of urban construction and spatial form (Fig. 6.2).

The initial round of urban planning in Hefei was proposed in 1956. At that time, the "focal" pattern of the old city area was adopted in the urban spatial form, thereby enabling the city to expand naturally. The influence of the natural ecological factors enabled the new urban space to expand toward the east, north, and southwest. Accordingly, this expansion enabled the formation of the industrial warehouse district in the north, cultural educational district in the southwest, and several industrial districts. Thus, the primary fan-shape urban form was formed.

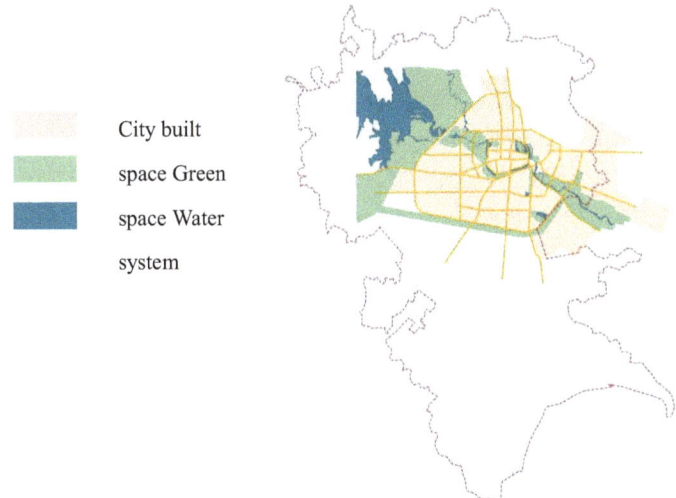

	City built
	space Green
	space Water
	system

Fig. 6.2 Schematic of the urban space of Hefei in 1956 (Painted by author)wl

The beltway was built in 1979. The old city area was the center of the total urban layout, while three industrial districts in the east, north, and southwest were established. Large vegetable fields and orchards were set among the industrial districts that eventually transformed into the public greenbelt. The limitation brought by the Dongpu Reservoir, Dafangying Reservoir, and Dashu Mountain in the northwest decided the development along the northeastern and southern areas of the city. An urban scenic district was built with the combination of the natural landscape of Dashu Mountain and the two reservoirs and the agricultural land. Moreover, the urban planning in 1986 just made a few minor adjustments (e.g., rebuilding of the Around-city Park) to the one in 1979 without fundamentally changing the urban spatial form (Fig. 6.3).

In 1995, unfavorable construction and construction-forbidden land were decided based on the natural conditions and land-use planning. The low-lying area at the lower reaches of Nanfeihe River (east to the Nanfeihe River bank, west to Ma'anshan Road, north to Tunxi Road, and south to the Chaohu polder area) was allotted as the urban ventilation passage because Dongpu Reservoir and Dafangying Reservoir are the first-level protection areas and key spillways in the city and construction-forbidden area. Moreover, accompanying the urban development in stages was the further release of construction land in the southeast, while the green wedge was further absorbed into the around-city greenbelt and the fan-structure was further strengthened. Additional space was likewise released in the northwest to foster ecological green wedge (Fig. 6.4).

From 1956 to 1995, the urban spatial form was an old city area-oriented around-city development motivated by natural waters, mountains, and greenbelts. The central

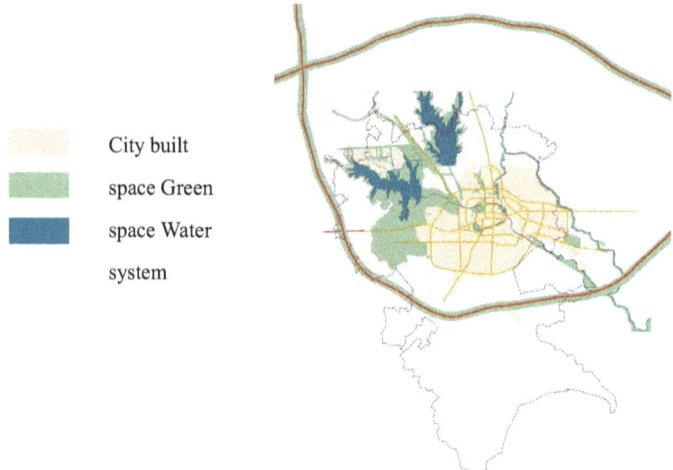

Fig. 6.3 Schematic of the urban space in Hefei in 1979 (Painted by author)

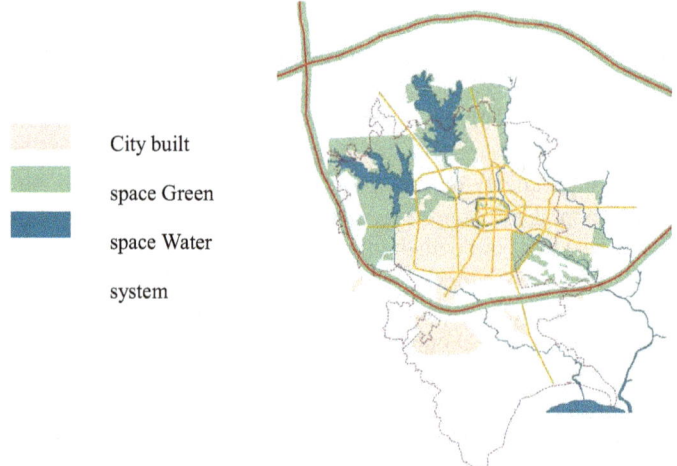

Fig. 6.4 Schematic of the urban space in Hefei in 1995 (Painted by author)

position of the old city area was highlighted from the original focal form to the fan-shape form. The urban spatial border expanded as well with the development of the city and expansion of the city scale. The simple development of the old city area cannot afford the comprehensive urban spatial system in Hefei that an urgent need for the transferring from the "around-city period" to "around-lake period" was around the corner.

Around-Lake Period

The fifth urban planning in 2006 was a leap-forward development on Hefei urban planning history. The government made an adjustment on the planning thought and decided to construct "141" urban spatial system, thereby signifying the beginning of the implementation of the developing strategy of Hefei urban form structure from "around-city period" to "around-lake period." In comparing this round of urban planning with the last round, we can clearly determine the rapid expansion of the urban land use scale and the urban space. The development of land use remained focused on the southwestern and southeastern regions of the city because of the limited natural factors. The land-use situation in the north has changed minimally. Moreover, green wedge in the southeastern and northwestern areas was replaced by the urban built-up area. Hence, the fan-shape structure no longer exists, while the urban structure was transformed into a cluster structure (Fig. 6.5).

The Hefei urban development entered a rapid development period with the adjustment of the administrative division of Chaohu City in 2011. An "around-lake period" has arrived with the construction of the around-Chaohu Lake demonstration area, which considers the Chaohu water area as the ecological center and as the national ecological civilization demonstration area with the slogan of "A large city beside Chaohu Lake, a highland with innovation spirit." In 2013, the Hefei urban planning decided to expand the urban spatial structure as a "1331" form and set up the spatial structure of the "Two-center, two-fan and two-wing" in the main city area. "Two-center" refers to the "old city center" in the old city area and the "new city center" in Binhu district. In 2013, Hefei's total spatial form changed substantially from 2006. To maximize the resource advantage of Chaohu Lake and strengthen the close connection of the Chaohu and main urban areas, the urban built-up area expanded rapidly

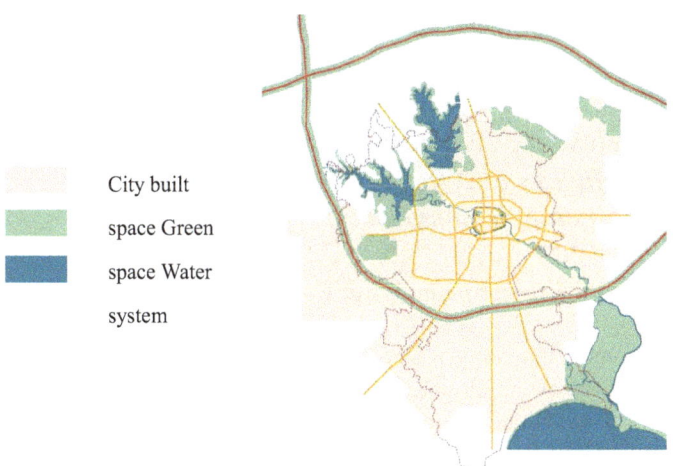

City built

space Green

space Water

system

Fig. 6.5 Schematic of the urban space in Hefei in 2006 (Painted by author)

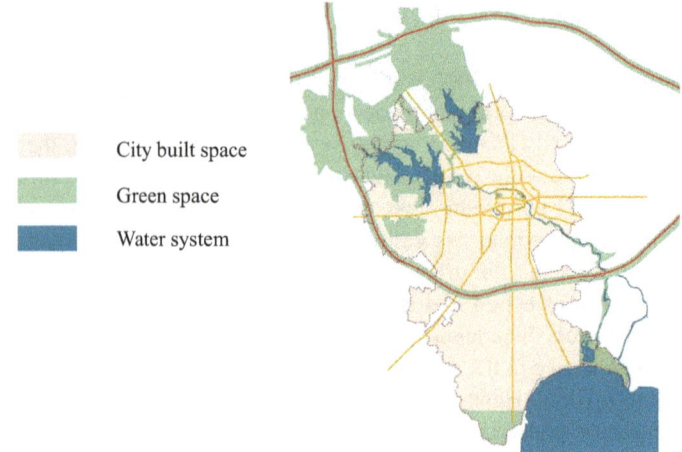

Fig. 6.6 Schematic of the urban space in Hefei in 2013 (Painted by author)

toward the southern part of the city and the northwestern and southeastern spatial border had minimal changes because of the limited waters and geology (Fig. 6.6).

The new Binhu District gradually came into being since the establishment of the "141" cluster form with the transformation from "around-city period" to "around-lake period" from "fan-shape" old-city-area-oriented urban layout to "141" cluster form, then to "1331" urban domain spatial strategy. Moreover, numerous urban clusters have been built up and the central position of old urban area has been weakened. With the incorporation of Chaohu Lake, Hefei entered the "around-lake period," while the "1331" spatial form is a natural extension of the "141" form. The establishment of a new city sub-center and ease of population and transportation pressure in the old urban area ensured the healthy development of the city (Aifeng & Yongchao, 2015).

6.2.4 Conclusion of Evolution Principle of Urban Spatial Morphology in Hefei and the Analysis of Its Influencing Mechanism

Conclusion of the Urban Spatial Form Evolution Principle

The urban planning diagrams of Hefei over the years and summary and conclusion of the spatial form evolution process has yielded the three stages of "focal form–fan-shape form–conurbation form" in terms of the evolution of the urban spatial morphology of Hefei.

(1)   1956–1979. The urban spatial morphology in Hefei was developed in a ring structure that is focused on the old urban area during this period. The cohesive effects of the urban center indicate that the urban space showed a naturally developed small focal form.

(2)   1979–1995. Around-city Park was built during this period in the main urban area. In particular, the surrounding green wedge gradually permeated into the around-city greenbelt to release the construction land in the southeastern area of the city. The urban spatial form during this period was a fan-shape form centered on the old urban area with expansion toward three wings (i.e., east, north, and southwest).

(3)   1995–2013. During this period, the urban spatial morphology in Hefei began its cluster-formed development. From the ecological perspective, the Hefei urban construction is generally influenced by the four waterways of Dongpu Reservoir, Dafangying Reservoir, Nanfeihe River, and Chaohu Lake; the limitation of the Zipeng Mountain and Dashu Mountain; and the greening vegetation space toward urban spatial form. In this period, the urban built-up space of the northwestern city had stopped its expansion. Meanwhile, the position of Chaohu improved rapidly within the city domain in the past years. The urban space expanded southward to maximize the resource advantage of Chaohu Lake and strengthen its connection with others. A conurbation urban spatial form with multiple centers was formed along the important axis (Song, 2004).

Influencing Mechanism of Urban Spatial Form Evolution

Urban construction is a multi-task undertaking that involves economic, political, social, ecological, and other factors. The change of the urban spatial form is an integrated output of various elements. The influencing mechanism of the evolution principle of urban spatial morphology in Hefei will be discussed based on the three aspects of economy, population, and transportation (Xiuli, 2010).

(1)   Economic promotion

Over the years, the economy is a core impetus in promoting the concentration and diffusion of a city. Given the superior location, abundant resources, convenient transportation, frequent business relations, and integrated development of service, technology, education, and culture, a strong attractive market is formed in the city through various economic activities. Accordingly, such market attracts the industrial elements to concentrate in the central city to generate an agglomeration effect. When the agglomeration effect improves to a certain level, a contagion effect appears in the area around the city. In the early period, Hefei was developed in the "focal" form because of the low social economic level and small urban built-up area. The reason is to weaken the influence of the natural factors of the surrounding waterways and mountains towards urban built-up area. Meanwhile, urban non-construction land and the natural spatial area were relatively large at that period, thereby forming the foundation for the transformation of large-scale green land into later green wedge and the formation of fan-shape urban form. Since the opening-up and reform, the urban economic level has been improved, the development and construction of industrial district has increased, high rise buildings in the city have begun to appear, and the urban scale has expanded rapidly. Given the construction of urban inner climate and urban ventilation passage and avoidance of the affection of industrial pollution towards urban built-up area, green buffer or isolation area were set in certain area around the industrial area to form a "three-leaf-fan" urban form (Liang et al., 2015).

(2)   Population support

The need for productive force has increased with the expansion of city scale, improvement of economic strength, and development of industrial scale. The promotion of urbanization attracts many peasant-workers into the city and the agglomeration effect of the city lead to the rapid increase of population in Hefei and the lack of residential resource and infrastructure. The urban space is pushed into the rural–urban fringe zone of Hefei. The original suburb area is absorbed gradually, new urban centers are formed, and the urban border expands continuously.

Therefore, the acceleration of urbanization in the rapidly developing Hefei introduces a stepped-up expansion of the city's urban space. (Yunfei, 2008).

(3)   Direction from transportation system

A city is the bearer of various material, capital, energy, and information flows within the city and between the city and outer space, while a full-order transportation facility can provide a springboard for it. An ancient city carried out the commercial activities through waterway transportation network. Hence, the urban structure was in axis-linear and fan-shape forms. Given the entry of the industrial society, the acceleration and perfection of transportation facilities promoted the internal and external communication between cities and the expanding speed of urban space. Hefei was renowned for its "fan-shape" urban form in history and its economic society has developed rapidly since the opening-up and reform. With the opening to traffic of high-speed railway, Hefei has formed an integrated land–sea–air transportation system centered on high speed railway, supported by highways and national and provincial roads, and affiliated by aviation and waterways. This ideal system directs the expansion of the urban spatial form of the entire city.

Therefore, we should make a comprehensive consideration of the diversified factors that influence urban spatial form evolution, make scientific reasonable planning based on specific characteristics of each city, and reasonably control the expansion of the city's outer border.

6.3   Deficiency and Outlook of the Construction of Urban Spatial Morphology in Hefei from the Ecological Perspective

6.3.1   Deficiency and Suggestion of Current Construction of Urban Spatial Morphology in Hefei

The conclusion and analysis of the evolution principle of urban spatial morphology in Hefei indicate that the current construction of the urban spatial morphology in the city has achieved periodical progress with a few deficiencies. Given the development of the city and the expansion of the city scale, Around-city Park has a certain controlling function of the old urban area. However, its influence is relatively weak on the level of the city domain and has minimal limitation on the expansion of the city. Meanwhile, the earlier "fan-shape" urban layout cannot limit the increase of the urban space because of the lack of attention to natural environment. The urban built-up area gradually replaces the natural space and causes serious problems, such as ecological environment pollution and poor living environment. With Binhu New District as example, the cyanobacteria phenomenon in Chaohu Lake appears annually. Although the relative departments have begun the treatment work, the degree should be strengthened.

Targeting the present deficiencies in Hefei's urban spatial construction, the corresponding constructive strategies are proposed from the macro, meso, and micro levels. Firs, a scientific strategic planning should be made at the macro level by considering the characteristics of Hefei, as well as other factors, and a reasonable allocation of various-usage land. Meanwhile, the concept of sustainable development should be considered to reserve sufficient land for the future development of the city. Second, urban construction in China can no longer adopt unordered expansion mode because of the acceleration of urbanization, high population intensity, limited resource share per capita, and lack of land resources. However, emphasis is given to the connotative development of urban space and urban form to realize the intensive usage of urban land to form a compact layout form. Lastly, sustainable development should be realized by enhancing the working strength of environmental management and improving the ecological environmental quality.

6.3.2   Developing Trend of Urban Spatial Morphology in Hefei

In terms of the development process over the years, Hefei has experienced the three stages of "focal form–fan-shape form–conurbation form." In comparing the present Hefei's urban construction situation with the urban planning diagram, we can clearly determine many contradictions in the original fan-shape urban form so that Hefei is carrying out the transformation from "141" to "1331" form. In the future, a dialectical consideration of Hefei's history and present developing situation indicates that the Hefei's urban construction should maximize its own resource advantage and make a scientific planning on future urban space in the suitable surrounding hinterland after considering the influence of all factors. Moreover, the urban spatial structure of "two-core and four-wing" is formed through continuous establishment and improvement of urban sub-centers and the full usage of abundant water resource and natural green developing elements to construct a new-type of cluster layout pattern.

6.4   Conclusion

Urban spatial planning is a long-term extensive undertaking. Urban spatial form is an important evidence for urban planning and construction and represents the interconnection of internal elements within the urban structure. This study concludes the principle of evolution of urban spatial morphology in Hefei through the summary and conclusion of the evolution process of urban spatial morphology in Hefei City and the influence of natural ecological elements toward urban spatial form. Moreover, this research analyzes the deficiencies based on the present situation of urban spatial morphology in Hefei and discusses the optimizing strategy and future developing trend from the ecological perspective, thereby providing guiding suggestions on urban spatial form to other cities.

References

Aifeng, P., & Yongchao, L. (2015). Introductive study on urban planning of Hefei towards evolution of spatial morphology since 1949. In *Annual National Planning Conference of China*.

Hefei Urban Planning Bureau. (1958, 1979, 1995, 2006, 2011). *Overall planning of Hefei city over the years*. Hefei Urban Planning and Design Institute.

Hefei Statistical Yearbook 2015 Editorial Board. (2016). *Hefei Statistical Yearbook 2015*. China Statistical Press.

Liang, H., Pei, L., Ling, Y., & Han, X. (2015). A study on evolution of urban spatial morphology and its driving forces—A case study on central urban area in Yangjiang City of Guangdong Province. *Planners*.

Qianqian, L., Yijun, L., & Wenyuan, N. (2011). A study of the relationship between urban spatial morphology and urban comprehensive strength. *China Population Resources and Environment*.

Qibiao, W. (2015). *The analysis of expansion and dynamic mechanism of urban spatial morphology*. Shandong Normal University.

Song, S. (2004). The analysis of dynamic mechanism about urban spatial structure evolution. *Urban Planning Forum*.

Xiuli, W. (2010). Research on the evolution of urban spatial morphology and the spatial expansion in Hefei. Hefei University of Technology.

Yunfei, Z. (2008). *Study on the urban space planning in Hefei*. Hefei University of Technology.

Chapter 7
Study of the Ecological Adaptive Mechanism of Traditional Human Settlements in Sichuan Tibetan Areas Based on a Cultural Perspective

Linglan Bi, Xuejin Liu, and Zhengjun Zhang

7.1 Introduction: Cultural Expression of Human Settlement Ecological Adaptation

The diverse spatial pattern of traditional human settlement and its poetic description of an "ideal residence" stems from the ecological adaptation of human beings to their local environment, without exception. Through practice and theory interaction, and refinement with past dynasties, the adaptation principles have developed into customs that direct the corresponding construction activities, which have even been deified into religious creeds and doctrines, eliminating the subjective superstitious dregs of ancient people. This mainly involves how the sustainable development "vitality" can be acquired in adapting the natural environment. Because the traditional human settlements in the Tibetan areas of Sichuan are built in a harsher natural environment with low technology, a dwelling culture with more unique local characteristics is generated. The content is established along with nature, and it stresses holy inviolability combined with religious belief. Its form is based on social production organizations, emphasizing delights and expectations in residences, and highlighting the sense of ceremony in interpersonal interaction, which renders it a unique cultural value in the construction of a spatial pattern and its construction process. Thus, a unique landscape of Sichuan Tibetan dwellings is created among the mountains and rivers that fully embodies the philosophy of a "poetic way of living."

Based on the national main ecological function positioning of the Sichuan Tibetan area, improving the human dwelling environment and carrying forward the essence of Sichuan Tibetan dwelling culture based on the native ecological condition has become an important path for local sustainable development. In the urbanization of the Sichuan Tibetan region over the past thirty years, the traditional track has been

L. Bi (✉) · X. Liu · Z. Zhang
School of Architecture and Design, Southwest Jiaotong University, No. 8 Teaching Building, Chengdu, Sichuan, China
e-mail: bilinglan@swjtu.edu.cn

© Springer Nature Switzerland AG 2021
W. Li et al. (eds.), *Human-Centered Urban Planning and Design in China: Volume I*, GeoJournal Library 129, https://doi.org/10.1007/978-3-030-83856-0_7

gradually abandoned with the utilization of modern human dwelling construction technology and a management system, driven by rapidly growing space demand. However, the simple imitation of the extensive construction and development mode of the inland area has resulted in the separation of construction activities from the primitive environment. The structures of human dwellings and landscape interactions have been damaged, further resulting in the degradation of the ecological environment and the excessive loss of natural landscape resources. The regional human dwelling culture is roughly used for "symbols," the foundation of the human settlement landscape is shaken, and the urban landscape is alienated. Therefore, it is of special significance for urban and rural construction to extract the ecological adaptation elements and mechanisms in human dwelling environment construction in the modern technical support system.

7.2   Adaptive Mechanism of the Regional Human Settlement Culture and Its Research Methods

Traditional regional human settlement culture is born in the process of the multi-system coordination of "nature, society, and economy". Traditional building space organizational structure shows the distillation of the response to "space demand". Its inner mechanism is the adaptation and transformation of the original environment based on certain construction technologies under established economic conditions and the social production organization mode. Different disciplines have different study focuses on regional human settlement culture.

The field of urban and rural planning and architecture was initially a study of the interaction mechanism between spatial form characteristics and traditional construction technology, for example, "the relationship between the structural form of a certain city and the spatial form characteristics of a certain dwelling." Then, the adaptability of construction technology (structure, material, construction technology, etc.) and spatial forms (building space combination, street space form, town site selection, etc.), with some elements of primary environments (climate, water system and terrain), were studied, such as "research on the spatial form of a traditional street lane based on climate adaptability". The basic research methods included spatial typology, empirical simulation and efficiency analysis combined with surveying, mapping, geographic information systems and empirical monitoring and testing.

The study of sociology mainly starts from two perspectives: social organization structure and cultural origin. The former focuses on the study of the influence mechanism of the established human and community relationships on the spatial structure, in fact, it is the allocation of the space resources according to social laws. The latter is based on various manifestations of cultural phenomena (including all kinds of nonmaterial folklore activities, folk systems and rules, including material architectural symbols, space forms, etc.) to explore interactive relationships and trace the evolution of the human culture itself. Its research methods are mainly based on literature collection, field social surveys, interviews, observation analysis and logical reasoning. In recent years, the cross-interaction between various fields has prompted researchers to explore the deep mechanism of the formation of regional human settlement culture based on the summing up of their respective fields. More and more subjects have been included in research in this field. For example, from the ecological point of view, the principle of ecological adaptability explains the internal spatial organization pattern of traditional human settlements and the underlying reasons for landscape construction characteristics.

In the field of urban and rural planning and architecture, through investigation, surveying, and mapping, supplemented by research, the existing research on the human settlement environment in Sichuan Tibetan areas has focused on the analysis of its unique spatial morphological characteristics and its evolution process, as well as the impact of its social production and lifestyle on the morphological structure of human settlements. The field of sociology emphasizes the influence of multicultural changes (Benism, Buddhism, etc.) on the spatial patterns of urban areas. Therefore, although there are some articles that discuss the influence of regional natural environment elements on traditional human settlement space patterns, there are very few analyses of the original roots of human settlement culture. As a result, in the process of urban and rural construction in modern Sichuan Tibetan areas, the inheritance of new human settlements in traditional contexts often has its own form, but it loses its true nature. In order to explore the deep roots of the generation and evolution of Sichuan Tibetan traditional human settlement culture, our study sorted out the key ecological factors that could potentially affect the construction of human settlements and their adaptation mechanisms, starting with an analysis of the cultural elements in legends, folk customs, and customs of traditional human settlements construction. At the same time, by empirically simulating the cases of Malkang, Ganzi, and Jiajuzhai, we analyzed the mechanisms and use restrictions of sunshine environments, wind environments, topographies, and landforms on the evolution of human settlements.

7.3 Analysis of Traditional Habitat Culture, Natural Environment Elements, and Resource Endowments in the Sichuan Tibetan Area

Sichuan's Tibetan area is located in the border areas of the Tibetan and Han national cultures, and its human settlement culture is influenced by the interaction of the two cultures. The vast majority of local residents believe in Tibetan Buddhism. The site selection theory, position beliefs, and alpine worship in Tibetan Buddhist teachings and related cultures are reflected by the placement of temple, TuSi GuanZhai, and the town habitat. This is similar to the feng shui doctrine of the Han nationality, where, through mysterious and concealed judgments "Good fortune and Bad luck," it shows the respect of human settlement construction for the local hydrology, terrain and topography, geological environment, climate, and the needs for obtaining the necessary life and production materials. The construction mode of the towns along the ancient Tea Horse Road was affected by the construction technology in the Han area and the need for new town spaces due to the mutual market. In the process of interacting with the natural elements of Tibetan areas, these towns gradually formed their own characteristics.

7.3.1 Site Selection Theory in Tibetan Areas

Most Tibetan settlements in the early Tibetan areas were gradually developed around the temples and TuSi GuanZhai. These important facilities could only be constructed after the ceremonial site selection of the "request" and "review of land" by the Tibetan Buddhist monks in charge. Although the site selection ceremony is very solemn and mysterious, it is a necessary procedure that includes conducting field surveys on specific operations and checking the topography and outlook. In addition to the receding mountains, the undulating hills on both sides, the surrounding rivers, and the physiological comfort and psychological safety, the ideal "best place" requires the following conditions:

(1)   The distance between the two pastures should not be too far in order to exchange grazing during the winter and summer seasons.
(2)   Adequate building space.
(3)   Enough fertile lands for cultivation.
(4)   Abundant water to meet the needs of drink and irrigation.
(5)   Timber for building houses and wood for fuel.
(6)   There are stones that can be processed into grinding discs.
(7)   Stone is available for building houses.

   Because water quality has an important impact on human health, Tibetan Buddhist scriptures think the ideal water source should have eight qualities: sweetness, coolness, softness, lightness, cleanliness, lack of smell, lack of damage to the throat, and lack of harm to the abdomen. This kind of water is called the water of eight merits. These conditions cover the necessary resources for the construction of monasteries and towns that include water sources, land, production and living materials, and building materials. Under the teachings of Tibetan Buddhism, these are the specific requirements for natural elements when constructing monasteries and towns.

7.3.2 Bonism's Animism

The Bon religion is a native religion of the Tibetan people that is formed by the interaction process between Tibetans and nature. The natural environment in Tibet is complex and changeable, and disasters are frequent. Inhabitants here feel very small and powerless when facing nature and advocate the belief that "all things have spirit", and demonstrate a sense of awe of mountains, rivers, lakes, animals, plants, thunder, rain and other natural phenomena. Practitioners of Bonism even think that without the permission of deities, any nature-influencing activity may offend the gods and cause disaster and disease. Therefore, their construction activities are very prudent, and many related taboos are formed in the tradition. For example, people avoid breaking ground in areas where plants and creatures are flourishing; they do not cut and dig on a holy hill, etc. In general, this religion pursues "non-intervention" or "less intervention" as much as possible in the environment. As a result, Tibetan people generally do not change terrain. Given that each site is "borrowed from the god of land," there are few things that will damage the site and the environment. The residential settlement respond to the mountain and the heights are scattered. The entire settlement is in harmony with the environment, which is essentially attributable to the extremely fragile ecological environment and the scarcity of suitable land resources.

7.3.3 High Mountain Worship

Tibetan residents generally believe the mountain is a ladder that enables people to go into and out of heaven, and thus, came to establish mountain-related beliefs and worship. Likewise, in the Tibetan areas in Sichuan, the holy mountain plays an important role in daily life. In addition to the pilgrimage to the mountain, the sites and orientations of monasteries, local officials, and residences are closely related to the holy mountain. As is described in Mkas-pavi-dgav-ston,[1] the situation of the age of Teras in Tibet was like this: "A li, which was located in the north of Tibetan, was surrounded by three gorges like a pond. The central Weizang was shaped like a ditch. The southern Duo Kang resembled farmland. All of them were submerged in the sea at that time".[2] The annals of history, the master work of Tibet, has a similar record: "Tibet was flooded by water at first." Therefore, the location of the Tibetan people's residence is inevitably subject to the hydrological condition, as most of them live on the higher ground, just as described in the book The Holy Martial Art.[3] In the late Qing Dynasty, Wei Yuan writes that "The 18 cities under the jurisdiction of the entire Tibet… The residences and the palaces are built cut into the mountain." This situation is also common in Tibetan areas of the Sichuan province. Additionally, in some areas of the Tibetan area in Sichuan Province (such as the Jiaju Tibetan village), local people attach great importance to the possibility of seeing the mountains when opening windows. This is because the terrain is so steep that houses must be built at

---

[1] The book was written in the 1660s; it was an important historical book describing the overall situation of Tibet.

[2] Hao (1980).

[3] Yuan (1984).

a certain height to see the holy mountain. Accordingly, the residents avoid the lack of sunlight by living at the bottom of the valley.

Field reconnaissance, folk customs of the living environment, and a literature review of Sichuan Tibetan areas show that the regional traditional residential settlements (including towns) are limited by natural conditions (terrain, topography, hydrology, etc.). The scarcity of constructible land has a clear impact on the shape of settlements and site selections. Firstly, the settlements are mostly close to the river. If the two rivers meet, they are closer to the bend, and if the river is too wide, the settlements will retreat to a certain distance and will be built on higher ground to prevent water damage will still getting the benefit of water. Secondly, because of the complex geological conditions, fragile ecology and vulnerability to geological disasters, settlements in the Tibetan areas of Sichuan, construction emphasize the adaptation to the topography, the protection of the original environment, and limited disturbance from construction. Lastly, because of the diverse climate and the long cold spells, settlement construction emphasizes access to sunlight and shelter from the wind. Essentially, the constraints of traditional residential settlement include topography, hydrology, sunshine and wind in the Tibetan areas of Sichuan. The specific details of how these ecological-environmental elements constrain construction should be demonstrated using actual conditions. For this study, selected settlements with historical extensions in the Tibetan area of Sichuan Province will be used as examples.

7.4   Empirical Analysis of the Ecological Adaptability of Traditional Human Settlements in the Tibetan Area of Sichuan

The Tibetan area of Sichuan is located at the southeast edge of the Qinghai-Tibet Plateau. Due to environmental conditions, development has lagged. By 2017, there were 107 cities and towns based on national standards.[4] The topography of the area varies greatly, and its shape is complex. According to the clustering analysis of topographical features, 94 cities and towns are distributed in high mountain valleys areas and 13 in plateau areas. Among the cities and towns, 51 were built after 1999, 34 were built after 1960, and only 9 were built before 1960.[5] Towns with long histories

---

[4] Including Ganzi Tibetan Autonomous Prefecture (hereafter referred to as Ganzi Prefecture), Aba Tibetan and Qiang Autonomous Prefecture (hereafter referred to as Aba Prefecture) and all towns in the Muli Tibetan Autonomous County of Liangshan Yi Autonomous Prefecture.

[5] According to Jiang Bin. *A study on the urbanization and social culture changes in Sichuan Tibetan regions—a case of Gengqing town in Dege County*, Liu Jiaoyan. *Sichuan Tibetan region's administrative regionalization from the twenty-fifth year of the Jiaqing's ruling era to the end of twentieth century* and Baidu Wikipedia.

and descriptions of the settlement's development and construction are very rare. Based on the survey and the data, this study selected three towns and one village as empirical cases for analysis.

7.4.1 Natural Environment and the Construction Process of the Case Study Urban Villages

Ganzi Town is a high-prototype town, located at the northeast edge of the Hengduan Mountains, upstream of the Yalong River, and in the transition zone between Qiuyuan and Shanyuan. Ganzi Town belongs to the continental monsoon climate zone, is a tribal pastoral area with flat terrain, fertile land, adequate water source, and rare farmland in Sichuan Tibetan area. The town is built around the Ganzi Temple, which was constructed in the seventeenth century.[6] Due to the worship of the holy mountain, the Ganzi Temple is located opposite the Valloj Snow Mountain. The earliest building was the Tusi Guanzhai (which is the government resident) and the Degon Buffalo Temple. To the southeast of the temple is the Tusi Guanzhai, which have become the dual core of the early settlements, most of the construction is built around these two important buildings. Due to the resource endowment and the benefits of temples, government resident, and transportation, the Ganzi Town area formed a settlement and then developed into a town, becoming a regional commercial and trade center and a distribution center. It has always been the political, economic, cultural center of the Ganzi North Road and the battleground of the military. In the Qing Dynasty, it became one of the three major towns in the Kang District (Fig. 7.1).

Gengqing Town is situated on the southeastern edge of the Qinghai Tibet Plateau with high geomorphic changes. Although it is located in the subtropical climate zone, it has obvious plateau climate characteristics due to topography and geomorphology. In the fifteenth century AD, the Dege family gained the right to rule the town because of its tribal marriage.[7] Because it was a "land of ten goodness"[8] with a perfect feng shui landscape, the Sakya temple "Tangjia stura hall" was built as a family temple in 1446 and the Dege Tusi Guanzhai was built nearby. With the growing power of Dege Tusi, for the Dege family's safety and dominance, the surrounding residents were forcibly moved to Qingqing Town. Later, due to the proximity of the old tea-trading

---

[6] "Ganzi" means white and beautiful. During the visit to the town of Ganzi, the fifth Dalai's disciple, Qu Ji Onpengcuo, saw that there were many white stones at the location of the Ganzi Temple. He believed that the area was filled with Buddhist light, and built a temple here, calling it the "Ganzi Zhaxi Snowball Noblin," which means a beautiful and elegant land. He built 13 Gru temples in the Hall Tribal region; Ganzi Temple is one of them.

[7] When Decker 35 Tazaxi Sangher, the Ridge Master of the more Gala area took a fancy to his daughter Denden and offered to marry her. Botha Zasisiah took advantage of this opportunity to demand that the area of the yak plough be used as a dowry, thus obtaining a large amount of land and population, and the town became the land of the Dege family.

[8] Jeb, a great auspicious successor at that time, predicted that the town was a "ten good place" for feng shui landforms.

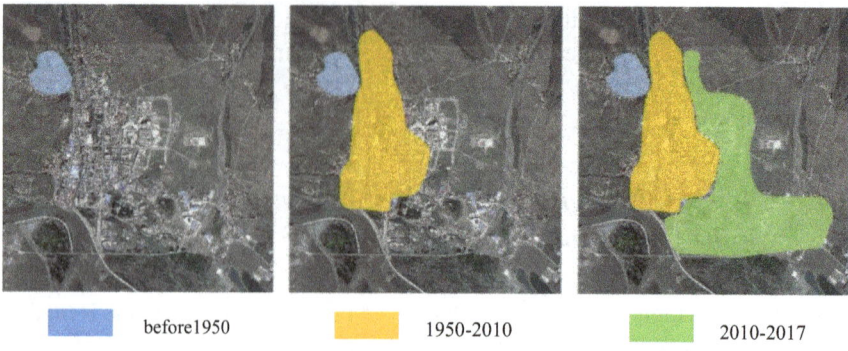

<table>
<tr><td>■ before1950</td><td>■ 1950-2010</td><td>■ 2010-2017</td></tr>
</table>

Fig. 7.1 Map of urban spatial evolution in Ganzi town (*Source* Prepared by author based on Google Maps)

route in the town of Qingqing, the transportation was convenient, and with sufficient cultivated land on both banks of the Oqu River and the Sequ River, the people gradually retreated to live on the banks of the Oqu River. In the seventeenth century, the Dege family built the larger Sakya temple "Lunzhuding Temple" (Jingqing Temple) and the new Tsi village. In 1729, the Dege Printing House was built on the west side of the new Tusi village. So far, three groups, called Sgengen, Nigend, and Ba Lunglong, have been formed around the temple and Guanzhai. Important buildings such as Guanzhai and monasteries are located among the residence instead of at the highest point of the settlement (Fig. 7.2).

Malkang Town is located in the middle of the Somo River and surrounded by mountains with complex terrain. Malkang Town has a continental plateau climate. In the south is Mount Changli, a subsidiary mountain of Mount Murdoch, and in the north is the Zanka Mountain in Chabei. Two hundred years ago, the earliest Bon temple in the Malcolm region was surrounded by Lacan. Due to the tribal disputes, the valley was full of danger and crisis. It was built on the Oryya Mountain at an altitude of 3400 m. At that time, there were temples and towns. In 400 AD, when Yuxing the living Buddha selected the site for the expansion of the Shenrab La Lacan, he found that at the intersection of the Somma River and the Dalangzu ditch and the foot of the holy Mount cha Kaban, the mountains, the river, the sky and the earth around the convex bank of the Somma River looked exactly like the symbol of Bonism-"卍" (Gyung Drung). Therefore, he thought there was "Gyung Drung in the sky, Gyung Drung in the earth, Gyung Drung in the river," and the place has a permanent and sturdy potential. Therefore, the "Five-star Burning Gyung Drung Temple" was built here. After that, the dependent temples gradually formed a settlement. Although the religious beliefs changed after 1880 and the Gyung Drung gling temples were finally converted into Tibetan Buddhism gelug Sect monasteries, it did not affect the congregation of believers and the development of the settlement. In modern times, due to the importance of geographical location, the development of trade and the gradual expansion of ~~the scale of~~ cities and towns, this area has become the seat of the Aba Prefecture government (Fig. 7.3).

Fig. 7.2 Schematic Map of main buildings and Residential areas in Gengqing town (*Source* Prepared by author based on Google Maps)

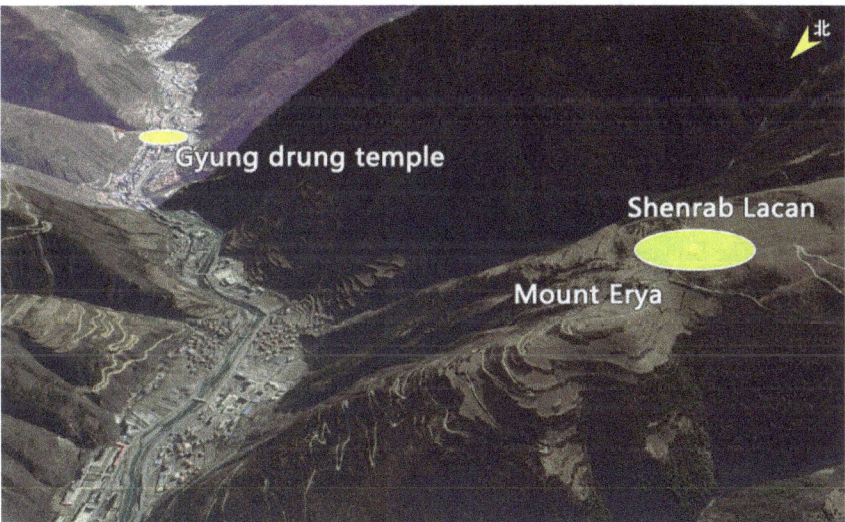

Fig. 7.3 The environmental conditions and location relationship between the Shenrab La can and the temple of Yongzhonglin (*Source* Prepared by author based on Google Earth)

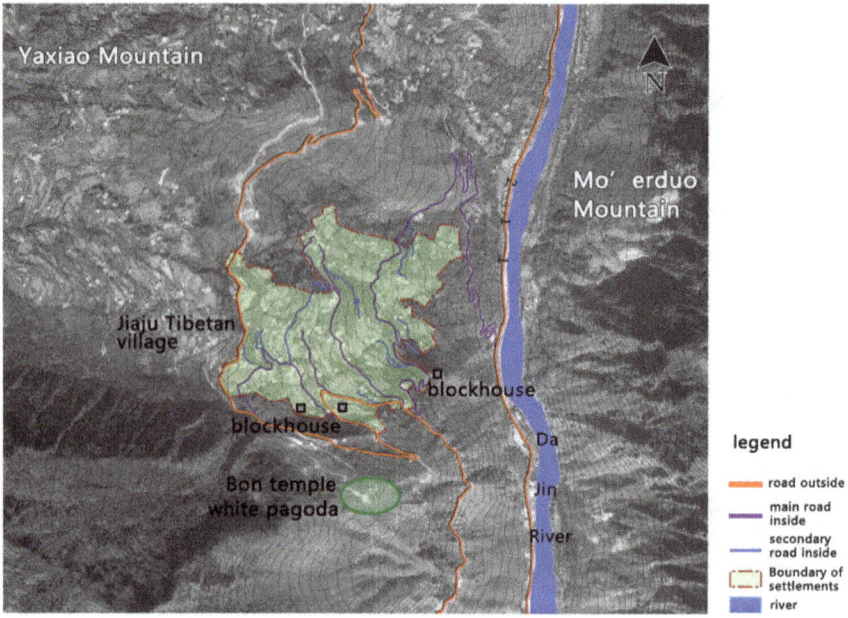

Fig. 7.4 The environmental condition and position of Jiaju Tibetan village (*Source* Prepared by author based on Google Maps)

Jiaju Tibetan village is located in the south of Niexia Township, Danba County, Ganzi Tibetan Autonomous Prefecture. The group's houses are scattered on the mountainside of the east-facing slope on the southeastern side of the Yaxiao Mountain. The earliest construction time has been unacceptable. The earliest mortuary building in the village was completed approximately in the middle of the Qing Dynasty. The Bon temples and white pagodas in the Tibetan village are located high on the south side of the settlement, opposite to Mo'erduo Mountain, and the inner houses of the village are facing Mo'erduo Mountain. The local folkloric "auspicious" home must be able to see the mountain (Fig. 7.4).

In the above cases, the development of important buildings (temples, Guanzhai) have all been relocated to the development of Ganzi Town, Gengqing Town, and Malkang Town. Therefore, in addition to the adaptive analysis of the ecological elements in a wide range, the case analysis will also focus on the adaptive changes before and after the relocation.

7.4.2 Adaptability Analysis of Terrain Elements

The Sichuan Tibetan area is in the mountain margin of the transitional area between the Qinghai-Tibet Plateau and the Sichuan Basin. Due to the complex topography, regardless of the specific elevation of the town, the slope of the construction land varies greatly. The overall slope is mostly between 8 and 45%, and the slope difference of the construction land in the same town goes up to 52° (Gengqing town). The data shows that the slope conditions of the temples and the Guanzhai land (mostly 8–15%) are generally better than the dwellings (mostly 15–35%). However, no temple, official residence or residential area occupies the flattest land (0–8%). These flat lands are often ploughed farmland closer to the river (Table 7.1, Figs. 7.5, 7.6, 7.7 and 7.8). Important buildings are often located at higher elevations of the settlement, backed by mountains or by civilian dwellings, overlooking river valleys and flat dams. Most of the settlements are located on the south and southeast slopes of the foothills. Due to the actual conditions, there are also some southwest slopes (Table 7.2, Figs. 7.5, 7.6, 7.7 and 7.8). The orientation of temples and Guanzhai is usually better than ordinary dwellings. This land is often able to receive solar energy earlier in the day and obtain enough sunlight in one day.

Most of the important settlements are located at the convex bank of the river intersection. Due to the difference in field topography and water potential, the key

Table 7.1 Slope statistics

The study area			Main slope (%)
Ganzi town	The old city	Overall plot of land	0–35
		Degongbufa temple	8–15
		Ganzi temple	25–35
		Mashu Tusi Guanzhai	8–15
		Konsa Tusi Guanzhai	0–8
		Resident settlement land	0–25
Gengqing town	Traditional settlement area	Overall plot of land	8–60
		Tang Jia sutra hall and the old Dege Tu Si Guanzhai	15–35
		The main building of the site selection construction in the later stage	8–25
		Resident settlement land	8–60
Barkam	The old town	Tibetan dwellings and their agricultural land	8–15
		Gyung Drung gling temple	Before relocation	15–25
			After relocation	35–45
Jiaju Tibetan village		Overall plot of land	8–45
		Temples and Bai Tallin	8–15
		Resident settlement land	8–15
		Terraced farmland	15–35
		Forest	35以上

Source Prepared by author

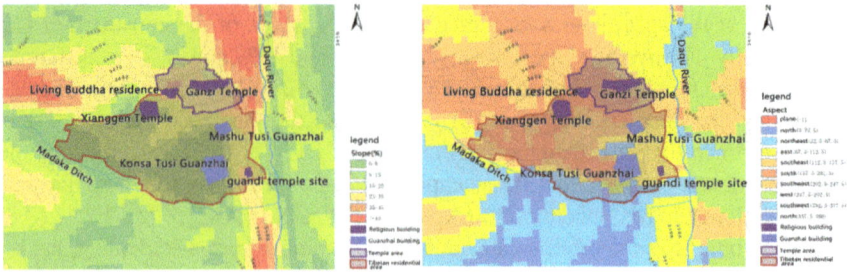

Fig. 7.5 Slope and aspect analysis of the Laocheng district in Ganzi County (*Source* Prepared by author)

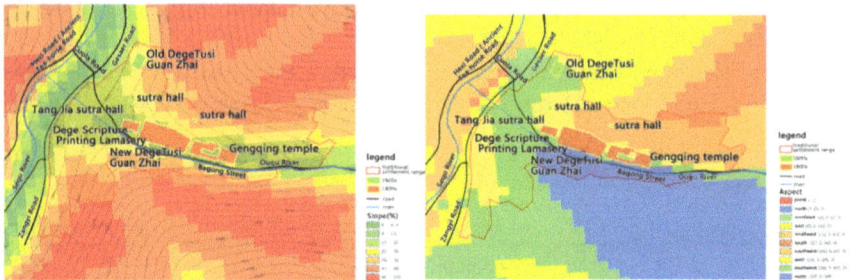

Fig. 7.6 Slope and aspect analysis of traditional settlement in Gengqing town (*Source* Prepared by author)

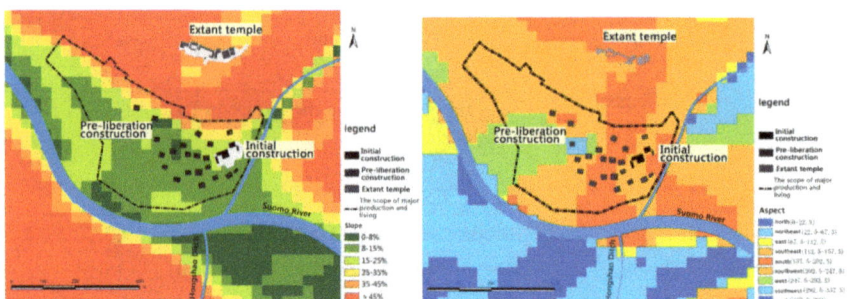

Fig. 7.7 Slope and aspect analysis of the main production and living area in Barkam (*Source* Prepared by author)

factors determining the location of the settlement are different, but the factors are mainly concentrated to the three aspects of "safety of settlement," "ease of life" and "improvement of irrigation." Among the settlements, Ganzi Town is located upstream of the Yalong River and the junction of the Daqu River. Due to the flat terrain, the Yalong River has a very gentle flow and a small drop (about 1.5‰). The river is overflowing, and the river channel is indeterminate. At present, the main

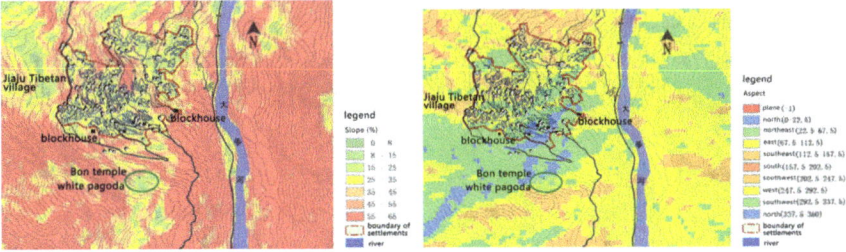

Fig. 7.8   Topographic slope analysis of Jiaju Tibetan village

river channel is about 200 m wide, 2000 m from Ganzi Temple and 120 m above sea level. The Daqu River flows into the Yalong River in this section, with a gentle flow, a relatively large ratio of 18 ‰ and a relatively fixed channel. The main river channel is 7 m wide, 170 m from the Ganzi Temple and 55 m above current sea level (Fig. 7.9). According to the historical data description, combined with the Ganzi hydrological data and a historical Google map analysis, the reason why Ganzi Town moved from the original location of the Degongbufa temple to the Ganzi Temple was related to flooding. The Yalong River has swung nearly 1000 m in recent years, and the situation in the junction area with the Daqu River is even more complicated. Degombufa shrine is too close to the confluence and is prone to flooding because of its low topography. The town of Gengqing is located at the junction of the two rivers. Among the rivers, the Sequ River is the main stream, the width of the river is about 30 m, and the range is 5000 to 14 per thousand, the water volume is large. The river is formed straight and slightly curved, slightly swaying and forming small patches along the river, which is the main agricultural land. The Ouqu River is a tributary, the width of the river is about 10 m, and the range is 5000 to 156 per thousand. The water volume is small, and the river form is straight. The early constructions of the Tang Jia sutra hall and the Dege Tusi Guanzhai were 100 and 130 m away from Sequ River, the height differences were 15 m and 20 m respectively, and they were 40 and 150 m from the Ouqu River. The late Gengqing temple and other buildings are close to the Ouqu River, but they are 250 m away from the Sequ River, and the elevation difference is more than 30 m. Because the two-river problem is not prominent, the location of the new Guanzhai and temples are more conducive to water intake. The town of Barkam is located at the junction of the Mosuo River, the Dalangjiao ditch, and the Hongshao ditch. The terrain and hydrological conditions are very complicated. The traditional settlement of Barkam is located at the intersection of Dalangjiao ditch and Mosuo River, and the right side of the estuary. Gyung Drung Temple near Dalangjiao ditch, and this temple is 150 m away from Mosuo River, and the elevation difference is about 20 m. Compared with the opposite Hongshao ditch, the two rivers of Somuo River and Dalangjiao are relatively mature, with smaller specific gravity, relatively slower water flow, and fertile soil. The area where the Jiaju Tibetan village is located is high mountain valley topography. The river is ferocious. Although it is located at the intersection of many rivers, the Jiaju Tibetan village is located on the gentle slope

Table 7.2 Aspect statistics

The study area				Main Slope direction	Remarks
		Overall aspect		Southward slope, 67.5°–292.5°	
		The temple area		Southward, 157.5°–202.5°	
		Tusi Guanzhai		Eastward and southeastward, 67.5°–157.5°	The slope of the land is small, so the influence of the slope is small
		Resident settlement land		Southward slope, 67.5°–247.5°	
Gengqing town	Traditional settlement area	Tang Jia sutra hall and the old Dege Tu Si Guanzhai		Northward slope, 22.5°–67.5°	
		Main building for later construction		Southward slope, 67.5°–247.5°	The GengQing Temple, the new Tusi Guanzhai, the Dege scripture printing Lamasery
		Resident settlement land	Southeastern part	Northward, 0°–22.5°	
			Southern part	Northward slope, 22.5°–67.5°	
			Eastern and western part	Southward slope, 67.5°–247.5°	
			Northern part	East–west slope, 22.5°–112.5°, 247.5°–337.5°	
Barkam	The old town	Tibetan dwellings and their agricultural land	Between southeast and southwest, 112.5°–247.5°		

(continued)

Table 7.2 (continued)

The study area				Main Slope direction	Remarks
		Gyung Drung temple (before relocation)	Southeastward, 112.5°–157.5°		
Jiaju Tibetan village		Overall aspect	Eastward, northeast and southeast, southwest, 22.5°–157.5°, 202.5°–247.5°	There are a small number of northwest slopes and north slopes
		Temples and Bai Tallin	Eastward, 67.5°–112.5°		

Source Prepared by author

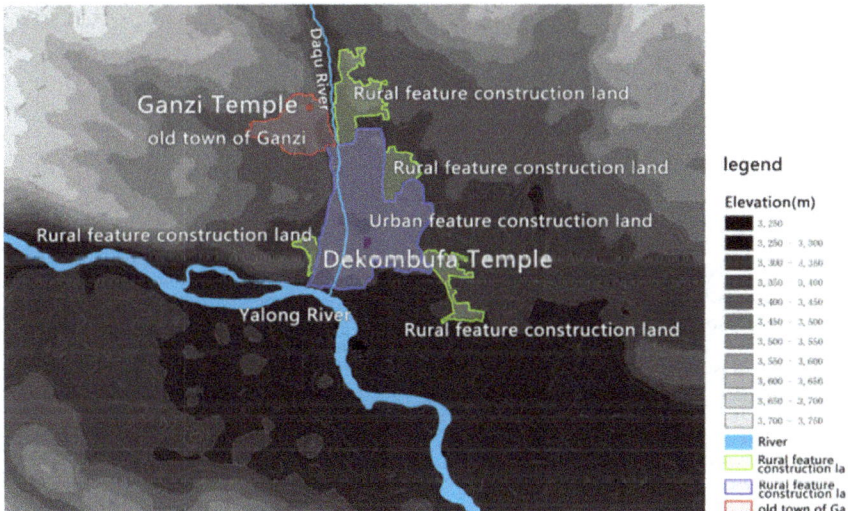

Fig. 7.9 Elevation analysis of the town area in Ganzi County (*Source* Prepared by author)

of the upper part of mountain and it avoids the impact of river disasters. Its daily water and irrigation use snow mountain melt water, and this water is introduced into the village through artificial ditches that are not related to the surrounding rivers.

Table 7.3 Sunlight conditions statistics

The study area			Winter solstice day		Summer solstice day	
			Sun duration (h)	Solar radiation (MH/m^2)	Sun duration (h)	Solar radiation (MH/m^2)
Ganzi town	The old city	Overall	9–10	2500–3500	10–15	7500–8500
		Resident settlement	9–10	2500–3500	13–14	7500–8500
		Ganzi temple	9–10	3000–3500	12–13	7500–8000
		Degongbufa temple	9–10	2500–3000	13–14	7500–8500
Gengqing town	Traditional settlement area	Overall	4–9	500–3500	7–12	6000–8000
		Tang Jia sutra hall and the old Dege Tu Si Guanzhai	7–8	1500–2500	9–11	7000–7500
		The GengQing temple, the new Tusi Guanzhai, and the Dege scripture printing lamasery	7–9	2000–3500	10–11	7500–8000
Barkam	Traditional settlement area	Overall	5–6	500–4500	12–13	3500–8500
		Gyung Drung Temple	6–7	2000–2500	11–12	7000–7500
Jiaju Tibetan village	Resident settlement		5–7	1000–2500	9–11	6000–7000
	Temples and Guan Zhai		6–7	2000–3500	10–11	6500–7000

Source Prepared by author

7.4.3 Adaptability Analysis of Climate Factors

This study analyzes the sunshine conditions[9] and wind environment of the above four towns, the villages, the winter solstice, and the summer solstice. The focus is on the differences between settlements and surrounding areas as well as important buildings and ordinary dwellings in settlements, as shown by the data.

Most of the land in the towns and villages have good sunlight conditions. Table 7.3 shows the specific data. The overall sunlight condition of Ganzi Town is relatively balanced, but the surrounding areas of Ganzi Temple have better sunshine conditions

---

[9] Sunshine conditions will be meausured by sun duration in hours and the amount of solar radiation (MH/m^2).

than other areas of the settlement, with more sunlight in the winter and less in summer. In Gengqing town, the level of sunshine varies greatly due to the topographical conditions. However, the Tangjia sutra hall, the old Tusi Guanzhai and the later construction of the Gengqing Temple, the new Tusi Guanzhai, and the Scripture Printing Lamasery have the best sunslight conditions in the area. Especially in the later temples and Guanzhai, sunlight is more abundant than it was for the early site selection. In general, having better access to sunlight is one of the reasons for relocation. The winter sunshine in the place where the Gyung Drung Temple in Barkam is located is longer than the average settlement time, but the summer is short, so it is possible to make full use of the sunshine and to avoid disadvantages. Due to the scattered layout of the residential buildings in the Jiaju Tibetan village, there are large differences in the sunshine conditions. There is more sunshine in the northwestern part of the settlement, and relatively less in the southwest and the east, but in general, these are in better locations in the region. The location of the temples in the Jiaju Tibetan village and the White Pagoda was at a higher position, so it spent more time in sunshine than the settlement and received more sunlight (Figs. 7.10, 7.11, 7.12, 7.13, 7.14, 7.15, 7.16 and 7.17).

The highest frequency of the average maximum wind speed and wind direction in Ganzi Town is the westerly wind. Combined with the topography, the wind direction

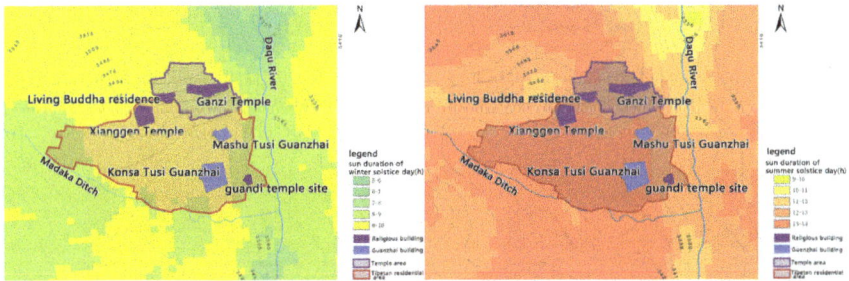

Fig. 7.10 Simulation of the sun duration of a winter solstice day and a summer solstice day in Laocheng district, Ganzi County (*Source* Prepared by author)

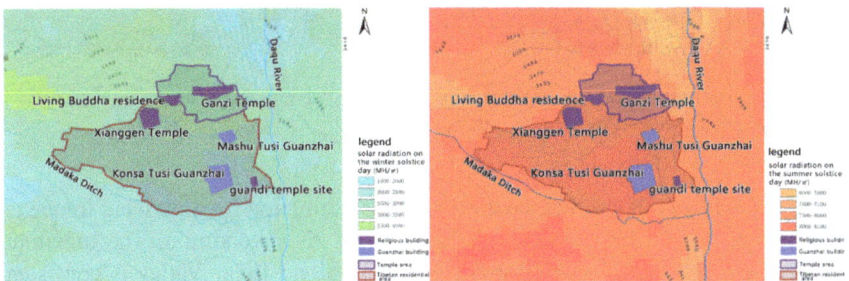

Fig. 7.11 Simulation of the solar radiation on a winter solstice day and a summer solstice day in Laocheng district, Ganzi County (*Source* Prepared by author)

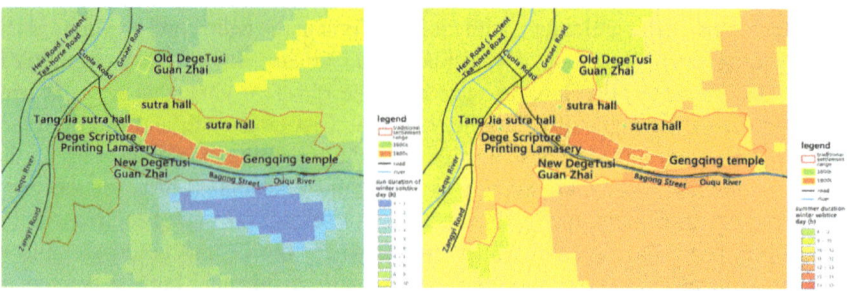

Fig. 7.12 Simulation of the sun duration of a winter solstice day and a summer solstice day in the traditional settlement area of Gengqing town (*Source* Prepared by author)

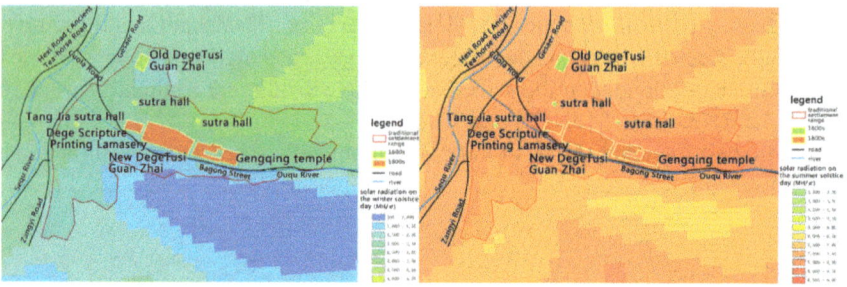

Fig. 7.13 Simulation of the solar radiation on a winter solstice day and a summer solstice day in the traditional settlement area of Gengqing town (*Source* Prepared by author)

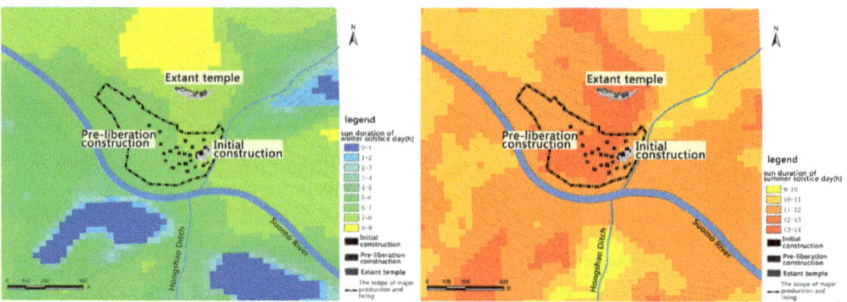

Fig. 7.14 Simulation of the sun duration of a winter solstice day and a summer solstice day in Barkam old town (*Source* Prepared by author)

frequency analysis shows that the dominant wind direction for high wind frequency is along the river valley, which is related to the river condition. The maximum frequency of the westerly wind and the broad Yalong River Valley is consistent, and the northerly wind with a higher wind frequency is consistent with the relatively narrow Daqu River. The old town of Ganzi is located between the two foothills of

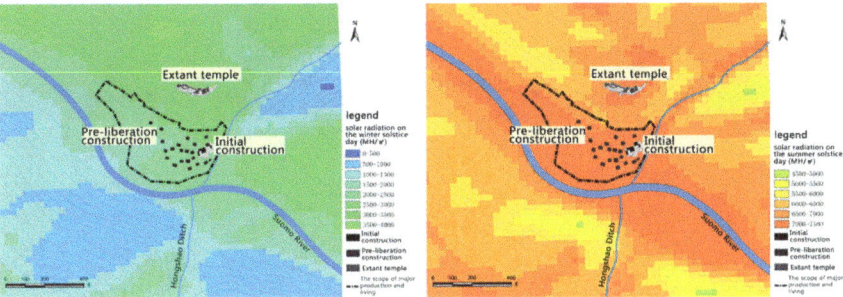

Fig. 7.15 Simulation of the solar radiation on a wintr solstice day and a summer solstice day in Barkam old town (*Source* Prepared by author)

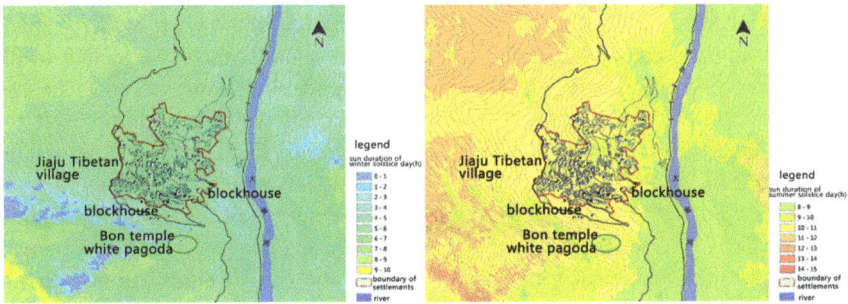

Fig. 7.16 Simulation of the sun duration of a winter solstice day and a summer solstice day in the Jiaju Tibetan village (*Source* Prepared by author)

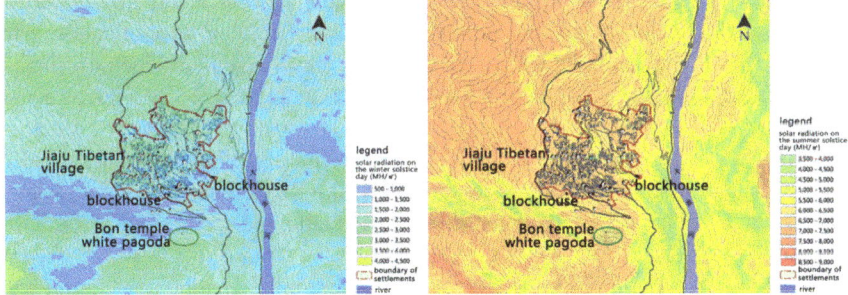

Fig. 7.17 Simulation of the solar radiation on a winter solstice day and a summer solstice day the Jiaju Tibetan village (*Source* Prepared by author)

the northwestern mountain range, avoiding most of the strong winds. The simulation of the wind environment in Ganzi town shows that the terrain is relatively flat, and

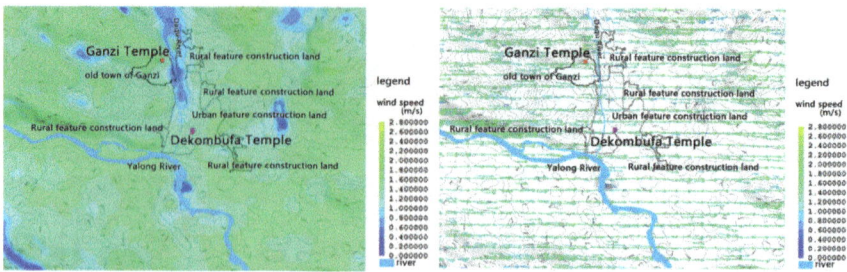

Fig. 7.18 Simulation of the wind environment in Ganzi town (*Source* Prepared by author)

greatly influenced by the dominant wind direction. Even under the most unfavorable conditions,[10] Ganzi Temple and the old city are located on the leeward slope with a small wind speed. The dominant wind direction in the town of Gengqing is southerly to the west, and the wind direction is essentially the same as that of the Sequ River. The windy season is mainly in the winter and spring seasons. Through the simulation of the Gengqing wind environment, the wind speed in the east, the south and the west of Genqing is high, the wind speed in Ouqu valley is low, and the wind speed of the Sequ valley is high. The traditional settlement areas are located where the wind speed is relatively small. The wind speeds of the Gengqing Temple and the new Tusi Guanzhai area built in the middle and late stages of the settlement are significantly smaller than the wind speeds of the Tangjia sutra hall and the old Tusi Guanzhai area built in the early settlements. The dominant wind direction in the area of Barkam is the northwest wind, which is consistent with the Mosuo Valley. The higher the altitude, the higher the wind speed, and the wind speed in the valley is relatively low. The traditional settlement area is located in a relatively low wind speed area, in which the location of the Gyung Drung Temple is near the estuary of Dalangjiao ditch, and the wind direction is slightly deflected to avoid the dominant wind direction. The dominant wind direction in the area where Jiaju Tibetan village is located is the northerly wind, and the five rivers meet in the area.[11] The wind direction of the area is clearly affected by the river valley. The wind environment simulation analysis shows that due to the blocking and weakening of the cold winter wind in the eastern part of the Mo'erduo Mountain, the Dajin River changes, and most of the Jiaju Tibetan village is located in a place with a small wind speed (Figs. 7.18, 7.19, 7.20 and 7.21).

To summarize, the Sichuan Tibetan area is an extremely cold climate zone with complex topographical conditions. Each settlement is based on the river development and topography of the location, and each settlement chooses a relatively safer,

[10] In January, the negative effect of wind speed on the lowest mean temperature of Ganzi town was the highest. Its windward slope had a high wind speed, and the direction of the wind was changed by the slope of the hillside.

[11] In the vicinity of the Jiaju Tibetan village, there is the Xiaojin River in the northwest and southwest, the Dajin River in the north and south, the Gesheza River in the northeast and southeast, the Donggu River in the southeast and northeast, and the Dadu River in the northwest and southeast.

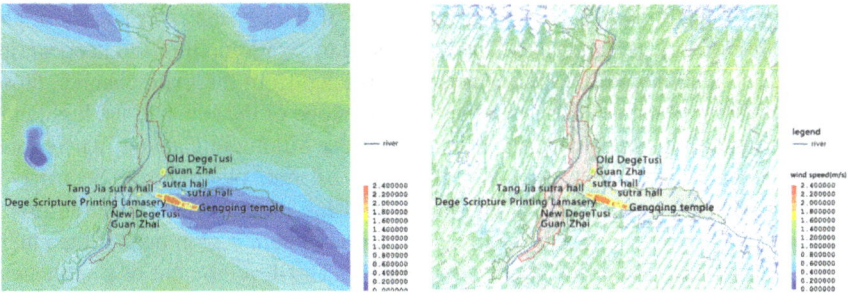

Fig. 7.19 Simulation of the wind environment in Gengqing town (*Source* Prepared by author)

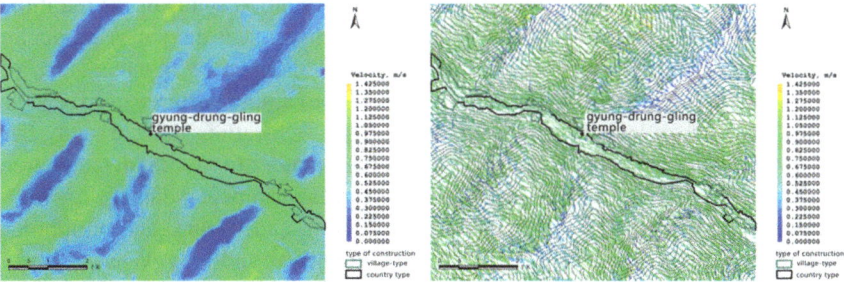

Fig. 7.20 Simulation of the wind environment in Barkam town (*Source* Prepared by author)

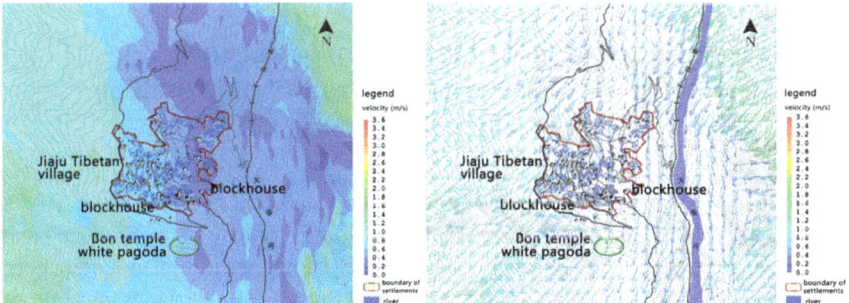

Fig. 7.21 Simulation of the wind environment in Jiaju Tibetan village (*Source* Prepared by author)

more convenient water-based area that does not occupy the cultivated land. Due to the shelter from the sun in the winter and the comfort of the summer sun is very critical to the comfort-index of the living environment. Based on the climatic factors such as temperature and precipitation, combined with an analysis of the development of the four settlements, it was found that the site for settlements were focused on better access to sunshine and protection from severe wind conditions. The important temples and buildings in the settlements are often located in places with better above conditions. Ganzi Town is located at the source of the plateau river and the high-altitude continental climate zone. Due to the need to avoid flooding and keep warm, the traditional settlements were located at a distance from the main channel, close to the stable tributaries of the wind, the southern slope of the sun, and the highlands on the southeast slope. Gengqing is located in the Mountain Canyon, where land resources are very rare. In order to save land, the larger plains of the Sequ River are reserved for farmland, and valley slopes are using for construction in traditional settlements. To prevent secondary disasters, the mountain region behind the chieftains' official does not allow grazing and cutting. The development of a settlement shows that the construction of a backward period will develop towards a region that is more conducive to water intake and shelter from the sun, tending to strive for more sunshine, and avoiding the dominant wind direction to prevent temperature loss due to wind and winter. The spatial relocation and construction status of the traditional settlements in the town of Malkan indicate that changes in the social production patterns have caused changes in the interaction of settlements with natural environmental conditions, and settlements have moved from alpine pastures to river valley farms. The later settlements emphasize the selection of locations that are sunny, have easy to use water, and are less subject to flooding. In order to save land, temples and dwellings choose more slopes. In order to prevent environmental damage caused by construction, there are strict regulations for hunting and fishing, and prohibitions on picking, logging, and grazing in the holy mountains and holy waters. The folk customs of "Watching the Sacred Mountain" in the Jiaju Tibetan Village are closely related to those of the Tibetan houses.

7.5 Ecological Adaptation Mechanism of Location Selection in the Heritage of the Traditional Dwelling Culture in Sichuan Tibetan Regions

The Sichuan Tibetan regions are located on the southeastern edge of the Tibetan Plateau, which is the transitional area from the Tibetan Plateau to the Sichuan basin. In these regions, landforms can be divided into two main types: a high mountain gorge region and a plateau region, among which local climate characteristics are different under the influence of the developmental forms of river valleys and altitudes. The influencing factors of the settlement location of the two regions are slightly different.

7.5.1 Adaptation Mechanism of the Land

There are also differences in the development of the water systems under both landform conditions, and for the specific location selection, water systems are taken into consideration, combined with safety, production, etc. In location selection, to prevent flooding, upland that is outside the swinging areas of the river or alongside tributaries with smaller swings is selected in the plateau regions. In the high mountain gorge regions, due to the rapid flow of the water and large fluctuations in the floods, the settlements are mostly on slopes of a certain height on the convex bank of the river. Additionally, at the intersection of the two rivers, settlements are often planned near stable tributaries. To facilitate water use, settlements are located as close as possible to the water source for safety and protection. For example, the distance from settlements to the great river is usually 200–300 m, and the vertical distance is 30–50 m, while the distance to a tributary is usually 30–100 m, and the vertical distance is within 10–30 m. In extreme cases, settlements can be planned on both sides of streams, or the water can be diverted into villages and towns.

Traditional settlements are rarely built on flat ground. They are built mostly in the 8–15% sloped area, the reason for which is perhaps that the 0–8% sloped land is primarily reserved for farmland due to the scarcity of suitable land for cultivation in Tibetan regions. In the high mountain gorge regions where the terrain is steep and the land is scarce, the settlements are mostly in the 15–25% sloped area in order to leave even more suitable farming land. Larger settlements are usually situated at the foot of the slope of the intersection area of two rivers or multiple rivers to use more alluvial flats as agricultural land.

There are few large-scale topographic reconstructions of the concrete construction of traditional settlements. Although the limited construction level is a possible reason, it is more likely due to the complex geological conditions and the fragile ecological environment, which may cause secondary geological disasters. Furthermore, the growth speed of organisms in a cold environment is slow, biological resources are precious in the cold environment. Thus, to ensure more abundant resources, settlement construction will avoid areas where bio-enrichment occurs, and accordingly, prohibited areas of construction, grazing, cutting, and picking will be stipulated.

7.5.2 Climate Adaptation Mechanism

The Sichuan Tibetan area is mostly located at high altitudes with typical cold climates, therefore, most tribes tend to settle on a sun-facing slope (southeast, south, southwest) because sunlight influences the microclimate and comfort of the tribes. In addition, to avoid the problem of shade from the surrounding mountains when living in the valleys, sites are often selected above a certain elevation. Furthermore, the local residents try their best to find places that can get more sunlight in winter and prevent excessive radiation in summer in the concrete construction process due to excessive high-altitude radiation. We can conjecture that the important temples and Guanzhai would move to sites with good climate conditions and reconstruct these places according to the actual development of the tribes.

In situations of arctic climate, wind has a great influence on microclimate comfort. Generally, tribe members pay attention to shelter when they select sites, especially the prevailing wind direction in winter and live mostly in no-wind and less-wind zones. Because it is windy at the top of a slope and there is less sunlight on the valley floor, tribes often settle halfway up a hill.

The constructions of Sichuan Tibetan traditional tribes are affected by the influence of many interconnected ecological elements, and the land interplays with the climate adaptation mechanism at different spatial scales and levels for the construction. This thesis emphasizes the action mechanism of the tribes and the environment from the perspectives of the macro and middle, so the tribes' concrete mechanisms in spatial organization and architectural scale will not be discussed further. Based on the natural environment in the region, people form fixed customs and taboos by constantly trying construction and relocation activities and accumulating experiences. The tribes of the ten good places fully embody the selecting principle that a dwelling place should be constructed with a close relationship between various living needs and natural resources. Empirical research indicates that the ten good places are closely related to the land's adaptability and the covering of various specific requirements, such as landform, geology, hydrology, and biology, in conjunction with construction taboos (land leases, bans on logging and grazing). The tribes' organizing principle of mountain worship embodies more of the climate adaptation when they select sites. If a Sichuan Tibetan can see the White Mountain during all seasons, it generally means the localities are above a certain elevation and at the south or southeast slope that provides more sunlight to cope with the cold climate and optimizes the microclimate for the tribe. Due to the many construction taboos derived from animism, site selections have two mechanisms: security and ecological suitability. Therefore, the organic combination of traditional tribes and landscape environment creates the unique Sichuan Tibetan habitat landscape under the low influence of the natural environment for tribal construction.

Acknowledgements The authors would like to the National Natural Science Foundation project of China entitled "Study on the ecological optimization mechanism of urban spatial structure in Sichuan Tibetan areas based on complex system theory (batch no.: 51578454)" for supporting this study.

References

Bo, C. (2005). *Urban bioclimatic design the approaches to realize the climatic rationality of urban spatial form*. Doctoral thesis. Retrieved from China National Knowledge Infrastructure.

Gao, H. (2013). *Traditional houses of the climate adaptation research*. Master's thesis. Retrieved from China National Knowledge Infrastructure.

Huang, G. (1980). Mkas-pavi-dgav-ston. *Excerpt* (4), 27–48+85.

Jin, B. (2016). *The influence of ecological factors on town's spactial distribution pattern in Sichuan Tibetan regions*. Master's thesis. Retrieved from China National Knowledge Infrastructure.

Jiang, B. (2003). *A study on the urbanization and social culture changes in Sichuan Tibetan regions—A case of Gengqing town in Dege county*. Doctoral thesis. Retrieved from China National Knowledge Infrastructure.

Liu, J. F. (1989). Tibetan ladder myth (4),93–94+87.

Liu, J. Y. (2011). *Sichuan Tibetan region's administrative regionalization from the twenty-fifth year of the Jiaqing's ruling era to the end of 20th century*. Master's thesis. Retrieved from China National Knowledge Infrastructure.

Long, Z. D. J. (2011). *Research on architectural culture of Tibetan Buddhism temple*. Doctoral thesis. Retrieved from China National Knowledge Infrastructure.

Tan, L. B. (2004). *Research on environmental evolution in traditional Dewlling houses*. Master's thesis. Retrieved from China National Knowledge Infrastructure.

Wang, X. (2014). *A study on form patterns of traditional settlements in central Shanxi from the view of environmental adaptability*. Doctoral thesis. Retrieved from China National Knowledge Infrastructure.

Wei, Y. (1984). *The holy martial art* (Vol V). Beijing: Zhonghua Book Company.

Wu, D. (2010). *The Tibetan traditional ecological ethics and realistic significance*. Master's thesis. Retrieved from China National Knowledge Infrastructure.

Xiao, J., & Yang, Y. Q. (2017). Landscape of traditional settlements in eastern Sichuan based on ecological adaptation. *Acta Ecologica Sinica., 13*, 4529–4537.

Yang, Y. W. (2013). *Magnet and the container: Sichuan Tibetan areas dege viewpoint of the formation and evolution of urban space form*. Master's thesis. Retrieved from China National Knowledge Infrastructure.

Chapter 8
Analysis of the Influence of Tibetan Buddhism on the Formation and Development of Urban Areas in Mongolia Region in Qing Dynasty

Chong Liu and Ying Han

8.1 Introduction

The evolution and development of human society formed human habitation, the city and town. Through the observation of different regional and historical layers, the formation and development of cities and towns reveal different characteristics and present different development schemes under the natural, political, economic and cultural influences. Due to limitations of being able to interact with the natural environment, ethnics living in the Mongolian area have formed an unique lifestyle that consists of no walls or rooms. They follow the water and grass while carrying yurts, herding and hunting as seasons pass.

After the Qing Dynasty began their reign over the Mongolian area, the ruler adopted the policy of "ruling by customs". He heavily advocated for Tibetan Buddhism so its growth could accelerate with a hope that it would tame the militant Mongolians by religious means. Soon after, temples rose up across the region along with an increasing number of monks. Under the co-effect of Tibetan Buddhism and the unique geographic environment of Mongolia, towns in this region during the Qing Dynasty formed an unique structure and development pattern as they were growing around the temples, such as Dorenol by Huizong Temple and Shanyin Temple and Xilinhot by Beizi Temple.

C. Liu · Y. Han (✉)
Inner Mongolia University of Technology, Hohhot, China
e-mail: hanywork2007@163.com

© Springer Nature Switzerland AG 2021
W. Li et al. (eds.), *Human-Centered Urban Planning and Design in China: Volume I*,
GeoJournal Library 129, https://doi.org/10.1007/978-3-030-83856-0_8

8.2   Reasons Behind the Formation of Towns

With support from the Qing Dynasty, Tibetan Buddhism became the major religion in Mongolia. The number of temples and the amount of religious activities in the area was increasing. Mongolian herdsmen were undertaking long, grueling treks to the temples for worshiping, which promoted the gathering of nomadic people living in Mongolia who before had no fixed settlements. These occurrences provided the basis for the development of commercial activities and consequently, regular trading markets began to form around temples.

Every spring during the Qing Dynasty, Jebtsundamba would hold the famous "Mandal" festival near Kulun where all affiliated herdsmen groups would gather. "Mandal" offered an opportunity for commercial trades. "The area was occupied by markets where vendors traded under temporary wood tents or from displays situated on the ground" (Gu, 1991). Later on, Lama opened a place for them to form a fixed market. Hexigten Banner's Jingpeng Town used to be a Lama temple without residents (Hua, 1916). But the market caused the area to thrive which helped form the town and a network of streets (Pozdenief, 1984). In the 23rd year of the reign of emperor Daoguang in Qing Dynasty (1843), the area had already reached the size and scale of a town (Jian, 1908). During the period of the reign of emperor Tongzhi in Qing Dynasty, the number of residents totaled 8,000 people (Hong, 2012).

The rise of temples also led to an influx of many handicraftsmen. As commercial activities expanded, handicraftsmen began to rent the temple's land to provide services for the daily activities of temples, religious ceremonies and visiting businessmen. Meanwhile, temples independently rented out large areas of farmland and grassland they received from rulers and nobles to further motivate the development of agriculture and husbandry. This helped form the urban space around the temple that featured the major industries of agriculture, husbandry, handicraft, and commerce Similar to Dorenol, after Emperor Kangxi built the Huizong Temple and Shanyin Temple, "the place was transformed from a dessert land without any food resources into a town"and further grew into "a busy commercial town with abundant Mongolian residents and crowded by visitors and buildings, becoming a metropolitan in the dessert" (Dong et al., 2008).

The essential activity of a city or town is the gathering of people. Temples helped attract a sufficient number of people to support the formation of a town. The thriving momentum caused by Tibetan Buddhism helped continuously cause cities and towns in Mongolian areas to develop.

8.3   Influence of Temple Location on Urban Development

Site selection is an important step before the construction of temples since location significantly influences urban development. Tibetan geomantic advocates animism. Additionally, the traditional mountain worship is the most primitive worship of

Fig. 8.1 The ideal layout

Tibetans. The Mongolian area sits on a plateau with an average altitude above 1,000 m. The terrain mostly features desserts and mountainous hills with rivers running through the valleys. Due to these geographic features, Tibetan Buddhism temples usually sit on top of mountains with gradual slopes in east and west directions, and they tend to face water sources. The site locations of temples provided good geographic conditions for urban development. From the geographical perspective, such a site location method is also scientific in adapting to the Mongolian geographic environment (Fig. 8.1).

Xilinhot was built around Beizi Temple and is located at the south foot of Aobao. The city is surrounded by hills to the east and west, has an open view to the south, and is bordered by the Aobao Mountain to the north. These geographic features help keep out cold winds in winter and provides the raw materials needed to build houses. In spring and summer, the ice and snow on the mountain melts and forms lakes at the plain before the town providing daily water resources along with the Xilinhot River for residents (e.g. Fig. 8.2). Uliastai and Dorenol also adopted very similar reasoning for their respective site selections (e.g. Figs. 8.3 and 8.4).

8.4 Influence of Temple's Spatial Layout on the Urban Spatial Layout

The original meaning of the Mandala is the "temple city" along with many other meanings such as "gathering" and "circular wheel". In Tibetan Tantric Buddhism, Mandala is not only the place and means for practitioners to communicate with gods, but also the ideal Buddhist state for believers. It represents the universal view of Buddhism. It is the belief that the world is centered on the Xumishan mountain where is surrounded by the four major continents, and the Tiewei mountain is the external discs around the world.

Fig. 8.2 Xilinhot (*Source* Author's drawing)

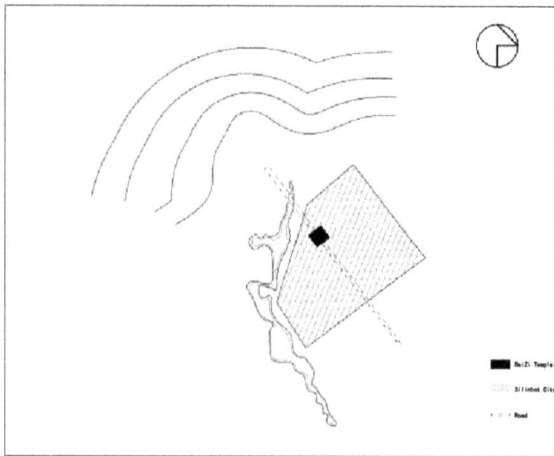

Fig. 8.3 Uliastai (*Source* Author's drawing)

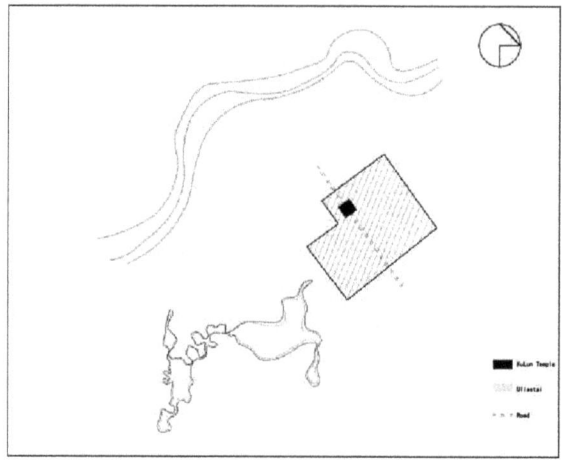

When constructing the temple space so it follows the Buddhist doctrine, the mandala-style spatial layout is often adopted to intuitively display the Buddhist universe. It also embodies the intrinsic property of "aggregation" and has a strong concentricity. Vajrayana is the foundation of Tibetan Buddhism and it follows the religious ceremonies and ceremonial order of "respect for worshipping"and "prayer to pilgrimages".Monks emphasize the merits of a temple tour since they pray and walk in the circular prayers forming prayer trail.

A town's layout is influenced by Tibetan Buddhists' performances of daily rituals and religious rituals centered around temples. Thus, temples became the center of the space, and towns grew organically around the center. The planar layout of the temple's ziggurat is transformed into an external prayer path, which further influences the spatial layout of the town. Kulun is an important town in the Monan Mongolian

Fig. 8.4 Dorenol (*Source* Author's drawing)

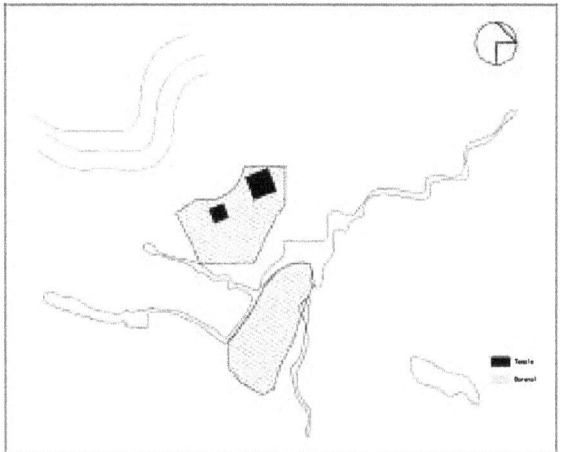

region. Early towns built residential areas and plazas around the Xingyuan, Xiangjia, and Fuyuan Temples in Kulun (e.g. Fig. 8.5). The towns of Drennol, Guihua, and others affected by Tibetan Buddhism share similar layout features in the early stages of town development (e.g. Fig. 8.6).

Fig. 8.5 Kulun (*Source* Author's drawing)

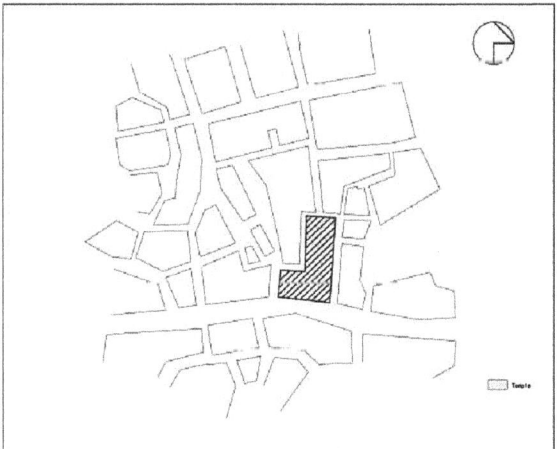

Fig. 8.6 Guihua (*Source* Author's drawing)

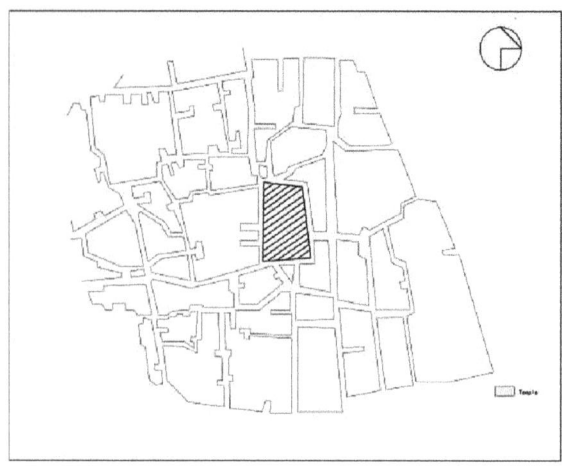

8.5 Development Patterns of Towns

8.5.1 Center—Multi-Core

Once towns started creating temples as the focal point, the initial town spatial layout was developed. As the size of the temple expanded, the number of monks increased and they began to build their houses near temples forming a clustered area around the centrally located temples. The layout of residential areas and squares around the temples were more flexible and free compared to the more strict layout of temples. For example, in the case of Guihua City during the Qing Dynasty, "Except for the Lama Temple that was higher than other buildings, people could only see a large number of houses and shops that were crowded together in disorder. The walls of the old city remained, but due to an increase in the number of residents in the city, many were forced to move outside of the city. Consequently, many new houses were built outside the city forming a large residential area" (Wu et al., 1993).

Since the lama needs a place to retreat, the temple restricted businessmen from opening shops near temples. Since businessmen could only set up tents in places far from the central temple area, they gradually built houses and opened branch offices or warehouses to form a "trade city". These areas became the second centers, thus forming a town pattern with dual-core clusters of "Temple City—Trading City". For example, Dorenol is a religious center and economic center in the desert region of Monan. It consists of a split temple area with two new and old trading sites divided by the Ere River. The two old and new temples located in the northwestern part of the city, Huizong Temple and Shanyin Temple, have a total of 13 living Buddha houses. In Dorenol's prime years, there were nearly a thousand houses the physical area of the temples were huge. Before the end of the Qing Dynasty, the size of the trading site in the southeast of Dorenol continued to expand. The tributaries of the Hutuo River crossed the city from south to north, and the two temples, Huizong Temple and Shanyin Temple, faced each other across the river to form a unique spatial and cultural pattern of "the north temple and the south residence".

8.5.2   Spatial Extension

Since most towns were formed spontaneously and without border walls, the streets led directly to the grasslands. Two gates or two opposite wooden pillars were usually used to designate the entrance of a town which formed unique prairie towns in Mongolia. The organic and free town layout provided a good foundation for the expansion of towns.

With the increase and expansion of the urban population and under the influence of natural geographical factors, towns used the temple as a starting point to grow at a certain axial direction. Due to the temperate continental monsoon climate in Mongolia, summers are short, and winters are long. Although the average annual temperature is low, winters are particularly cold. Additionally, the annual average precipitation is very low resulting in droughts and water shortages in most regions. In order to avoid the negative impacts of climate, the houses in towns were mainly oriented southwards and then continued to expand in this direction. The temples were the starting points of the axis and houses were arranged along the sides of the axis, so which eventually formed a main axis running through the entire town. Such as Xilinhot and Uliastai, Aobao Mountain and the temple on the north side of the town was taken as the starting point to form a north–south urban axis that led the entire urban space (e.g. Figs. 8.7 and 8.8).

Fig. 8.7   Xilinhot's city axis

Fig. 8.8 Uliastai's city axis

8.6   Conclusion

The rise and growth of cities and towns in the Mongolian area during the Qing Dynasty were significantly associated with Tibetan Buddhism which influenced the urban society and economy. To a certain extent Tibetan Buddhism motivated the economic development by gathering the population. Meanwhile, Additionally, the formation of urban space around the temple was influenced by Tibetan Buddhist spirits and and the geographical and cultural environment in Mongolia. By exploring the reasons, approaches, and types of the towns that formed and expanded during the Qing Dynasty, we have shown the importance for supporting the development, construction, and protection, of the existing Mongolian cities and towns.

Acknowledgements This work was supported by grants from the National Natural Science Foundation of China "Integrated Protection Based on the Value of Inner Mongolia Tibetan Buddhist Architectural Heritage System", project number: 51668049, and the Research Project of Inner Mongolia University of Technology "Research on the Relationship between Tibetan Buddhism and the Formation and Development of Tibetan Buddhism—Taking the Chahar Area as an Example", Project Number: X201613.

References

Dong, X. F., Zhang, P. J., Bai, L. Y., & Bai, X. (2008). An analysis of the locations of Tibetan Buddhist temples in inner Mongolia. In *The Academic Annual Meeting of China National Architecture Research Association and the Second Forum on the Development of the Protection of the National Architecture (Cultural Relics), Conference Proceedings.*

Gu, B. C. (1991). *Travel notes of Tibet.* China Tibetan Studies.

Hong, X. (2012). *Tibetan Buddhism.* Sichuan People's Publishing House.

Hua, Y. (1916). *Inner Mongolia minutes.*

Jian, H. S. (1908). Dorenol. *Oriental Magazine*, (10).

Pozdenief, A. (1984) *Mongolia and Mongolians*, vol. 2 (below). Inner Mongolia People Publishing House.

Wu, Y., Bi, L. G., Cheng, C. D., & Zhang, Y.J. (1993) *The general history of Mongolia.* Inner Mongolia University Press.

Chapter 9
A Method of Discovering Urban Functional Zones Based on POI Feature Vector and Network Kernel Density

Shiwei Shao, Hui Liu, and Lin Lu

9.1 Introduction

With the development of data-acquisition techniques, multi-source multi-scale vector data has become more readily available. However, the vector data generally only contains the spatial information of the elements and we can't directly obtain implied information, such as the classification results of different types of architectural construction in cities (Steiniger et al., 2008) and the extraction results of urban centers (Luscher & Weibel, 2013). Due to their advanced level of information, urban functional zones can create values that are consistent with their functions within a relatively limited area and spread superior capabilities to the surrounding areas. The division of urban functional zones lays the foundation for further urban spatial analysis, such as assessing the stability of the annual average temperature in each functional area of the city (Sun et al., 2013), etc. Therefore, the identification of urban functional zones has important significance in the verification of urban planning and land use planning (Xiang, 2014).

At present, there are four main types of methods for identifying urban functional zones. The first type is using remote sensing image processing methods to identify functional areas. For example, Jing (2008) established a model "remote sensing image interpreting semantic application" and based on urban development characteristics and classification of land use types. The second is the use of computer ways to identify urban functional zones. For example, Zhi (2010) calculated and generated a convex functional area using the Graham algorithm on the basis of Delaunay triangulation and triangulation clusters. The third uses statistics and probability theory to identify functional areas. For instance, (Yu et al., 2015) only took advantage of POI data to divide the central business district in the city applying network kernel density estimation methods. Chi et al. (2016) redistricted POI data, and the city was

S. Shao · H. Liu (✉) · L. Lu
Wuhan Natural Resources and Planning Information Center, Wuhan, China
e-mail: liuhuiwhu@126.com

© Springer Nature Switzerland AG 2021
W. Li et al. (eds.), *Human-Centered Urban Planning and Design in China: Volume I*,
GeoJournal Library 129, https://doi.org/10.1007/978-3-030-83856-0_9

divided into functional areas by constructing frequency density of indicator and type ratio. Kang et al. (2017) obtained POI density scores to identify urban functions of Wuhan City with the help of quantitative analysis of POIs point densities in various zones. The fourth type is using machine learning algorithms and models to identify functional areas. For example, Song et al. (2015) constructed a Gaussian mixture model based on the temporal and spatial distribution of pedestrian trajectory data, identified seven types of the urban land and matched the result with urban areas. Jiang (2016) designed a functional area recognition algorithm with multi-feature determinate weighting utilizing the Gaussian mixture model. Wang Yan et al. (2009) used K-Means to zone urban functional zones. Yuan et al. (2015) combined POI with taxi track data, to establish the topic model and used K-Means to perform regional aggregation to distinguish different functional areas. Shili et al. (2016) built latent Dirichlet model and Dirichlet polynomial regression models and used the OPTICS clustering results of different models and then obtained the recognition outcomes of urban functional zones. Gao et al. (2017) used Dirichlet's thematic model, K-Means, and Delaunay's triangular space constraint clustering method to extract functional areas in cities.

Yu et al. (2015) utilized fewer data types and a simpler network kernel density estimation method to divide the central business district, but it is limited to identify part of the urban functional zones. Other methods are more complex and require more data types. Therefore, this study is based on POI characteristics and road network constraints. K-Means method are used to divide the survey region into K functional areas, and then kernel density estimation and Kriging interpolation method are applied to further accurately identify the high density of various POIs in urban space in order to identify the multi-functional areas in the city.

9.2 Extraction of POI Feature Based on Map Segmentation

9.2.1 Map Segmentation

Road networks are usually made up of the city's main streets, such as express-ways and loops, which naturally divide the city into areas of different sizes. By transforming the urban vector model projection into a grid model, the urban road network can be converted into a binary image. After using the fuzzy morphological operators to process the image, a basic urban spatial unit is formed (Steiniger et al., 2008).Classical mathematical morphology is the use of set theory to achieve a series of nonlinear or linear changes in the image to reduce the effect of image noise (Luscher & Weibel, 2013). However, with complex information and strong correlation, the processing effect is not ideal. In order to solve the problems caused by the uncertainty and inaccuracy of image processing, some scholars have introduced fuzzy mathematics into morphology and have promoted the generation of fuzzy morphology (Steiniger et al., 2008). Fuzzy morphology deals with the problems of image processing and analysis by fuzzy set theory. In specific situations, it has better performance than traditional methods. Fuzzy set theory is a common method to deal with uncertainties, and is widely used in image, spatio-temporal data processing and other applications. Similar to traditional binary morphology (Luscher & Weibel, 2013), fuzzy morphology includes four operators such as fuzzy expansion, fuzzy corrosion, fuzzy opening, and fuzzy closing. Through the use of fuzzy operators, the urban road network can be accurately segmented to form mutually independent regions that do not cross each other, which can provide a new thinking for urban segmentation.

9.2.2 TF-IDF Extraction of POI Features

In order to obtain the semantic features of the POIs in each region, we use the TF-IDF algorithm, a statistical algorithm of word frequency of documents, to calculate the POI attribute of the road segmentation region (Xiang, 2014). TF-IDF is an algorithm for counting the word frequency of statistical documents. TF is used to describe the number of occurrences of topic keywords in the document (Luscher & Weibel, 2013; Steiniger et al., 2008), the more times, the higher the similarity between the keyword and the topic, and the IDF represents the inverse. The word frequency of a document is used to measure inimitable symbol of the topic keyword in all documents (Xiang, 2014; Zhi, 2010). The smaller the number of occurrences, the higher the recognition of the topic keywords in the document.

This paper uses TF-IDF to count the POI characteristics, and considers the area formed after the road network partition as a document. For each area r_i, i = 1,…, R, the POI category is regarded as a topic, and the POI category number is counted; then the characteristic attributes of a single POI are considered as term item, and then the TF-IDF of each type of POI in each region is calculated to measure the importance of the POI in the region. The TF-IDF value vij is calculated by Eq. 9.1.

$$v_{ij} = f_{tf} * f_{idf} = \frac{n_j}{N_i} * \lg\left(\frac{R}{\|\{count(r_j)\}\|} + 0.01\right) \tag{9.1}$$

where nj indicates the number of POIs belonging to the j-th category in the region i; Ni indicates the number of POIs in the region i; R shows the total number of regions, and count (r_j) indicates the number of regions including the j-th POI. The TF-IDF formula can easily be improved for situations where all regions contain j-type POIs by adding the harmonic value, which avoids zero-value events.

9.3   Identification and Verification of Functional Area

9.3.1   K-Means

The K-Means clustering algorithm is a typical partition-based method. It regards the number of divisions as a benchmark for data clustering (Zhi, 2010). The similarity of the same kind is higher. In contrast, the similarity of different types is lower. The basic assumption of the K-Means algorithm is: for the same cluster, the initial cluster center is randomly selected so that all the points in the cluster are less than the distance to the center of the other clusters (Yan et al., 2009). Given N samples set $X = \{x_1, \cdots, x_i, \cdots, x_N\}$, We divided them into K disjoint clusters, the sample mean of each cluster is μ_j, then in the same cluster the minimum mean square error must satisfy Eq. 9.2.

$$\sum_{i=0}^{N} \min_{uj \in C} \left(\left\| x_i - u_j \right\|^2 \right)$$ (9.2)

where N is the number of samples; xi is the i-th sample; u_j is the sample mean of the j-th cluster.

Using the K-Means, the regions divided by the road network are clustered according to the K value of the TF-TDF vector eigenvalues of the POI in the region, forming a K-class. The initial TF-IDF feature vector set is treated as the initial cluster number K, and the set is calculated according to Eq. 9.2. The whole sample space is divided into K regions by similar attributes. Through the distribution characteristics of POI data in Wuhan City, the cluster number is chosen as 8, and the clustering results are shown in Fig. 9.1. The 0th category indicates the areas without POI, and there is no sample data as a clustering result.

9.3.2   Network Kernel Density Estimation

Although spatial clustering results can reflect the distribution characteristics of POI data well in urban space, it does still not definitively determine the functional areas represented by the divided areas, such as mixed commercial and residential areas, which cannot be completely separated to independent parts. In order to further identify the urban functional zones, network kernel density is employed to identify high-value areas of various types of POIs in cities to accomplish urban functional zones discernment.

The kernel density estimation is a method that starts from the data sample and studies the distribution characteristics of the data itself without any assumption and any priori estimates (Li et al., 2004).The advantage is that it can calculate point strength anywhere in the entire study area (Wenhao & Tinghua, 2015). The network

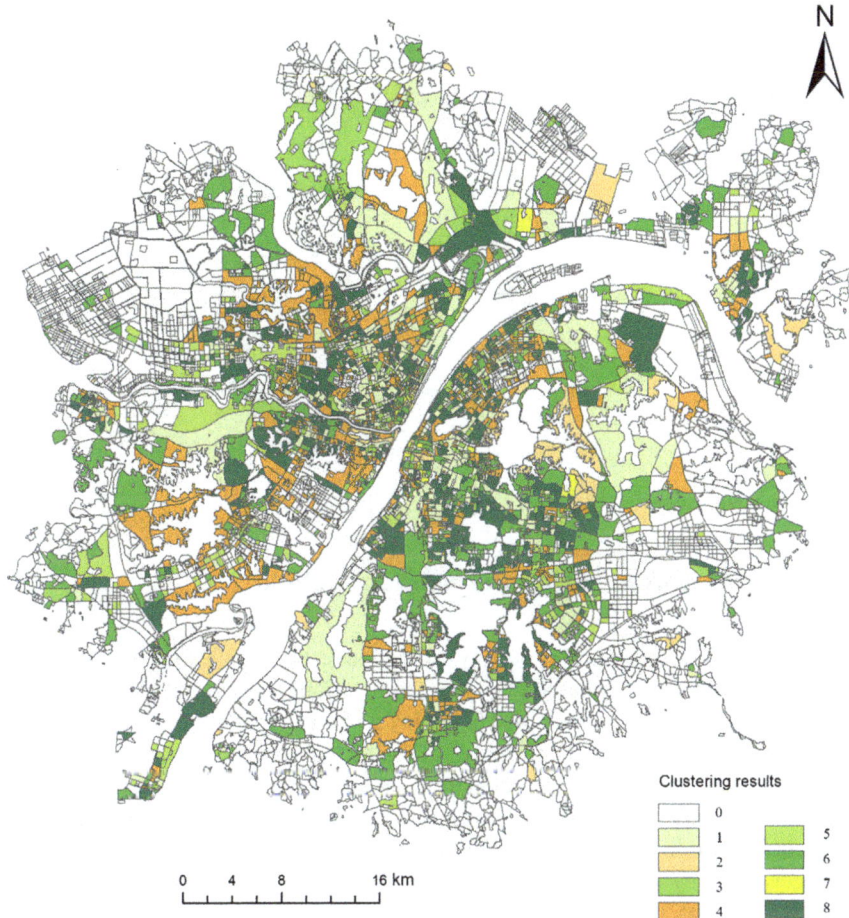

Fig. 9.1 Clustering result with 8 partitions

kernel density estimation can reflect the density distribution characteristics of the point element along the road network. The core parameter of network kernel density estimation is the distance bandwidth, which is mainly to explore the change of spatial distribution characteristics of geographical phenomena with various geospatial analysis scales. In general, the distance attenuation thresholds used in network kernel density estimation are 300, 600, 900, and 1200 m (Yu et al., 2015). Because of the "along roads" feature, the network kernel density estimation method is used to perform spot intensity estimation for each type of POI selected. Figure 9.2 shows the results of network kernel density estimation using the landscape category POI as a dataset and 600 m as the distance attenuation threshold.

Fig. 9.2 Network kernel density estimation based on scenic type of POI (600 m bandwidth)

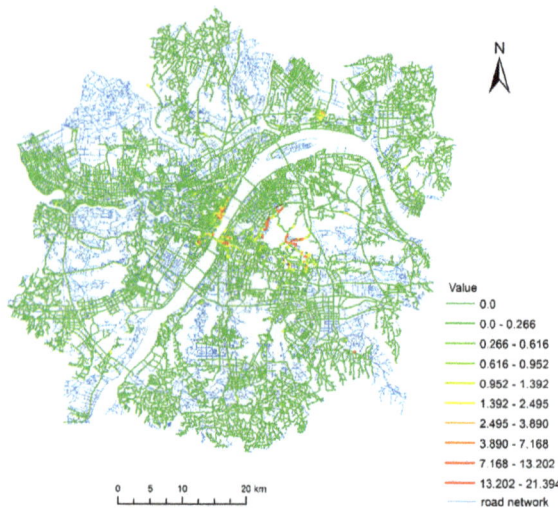

9.3.3 Spherical Kriging Interpolation Method

Since the above network kernel density estimation results are only shown on the road network, the division of the boundary of the functional area is not intuitive, so the method of spatial interpolation needs to play a part in correlating the results to the spatial range to be divided. The methods of spatial interpolation mainly include trend surface method and Kriging interpolation method. The fitting precision of the trend surface method is unique, and it does not require high fitting accuracy (Wenhao & Tinghua, 2015). In this study, it is necessary to divide the functional area by fitting data generated by interpolation, and the accuracy of the fitting data is demanding. Thus Kriging space interpolation method is accepted. The Kriging method includes the general Kriging method, the global cooperative kriging method, and other branch methods. The ordinary Kriging method is the most widely used. It assumes that the mean value of the data is unknown. Compared with the ordinary Kriging method, the global cooperative Kriging method assumes that the dominant data trend is known, and it is only used for qualitative analysis. Therefore, the ordinary Kriging method suits to spatially interpolate the set of points generated by the kernel density estimation. The various grams in the ordinary Kriging method are spherical models, exponential models, etc. The general spherical model is used for geo-statistical statistics, and the other is used for biological research. Therefore, we use the ordinary Kriging spherical model for spatial interpolation. We determine the corresponding search radius based on the different distance attenuation thresholds used in the previous network kernel density estimation. For example, if distance attenuation threshold of the network core density estimation is set to 600 m, the radius is set as same value, 600 m, applying ordinary Kriging spherical model to search for spatial interpolation. The Kriging interpolation results with the scenic spot type POI and the search radius of 600 m are shown in Fig. 9.3. By comparison, the K-Means clustering results are

Fig. 9.3 Kriging interpolation based on scenic type of POI (600 m search radius)

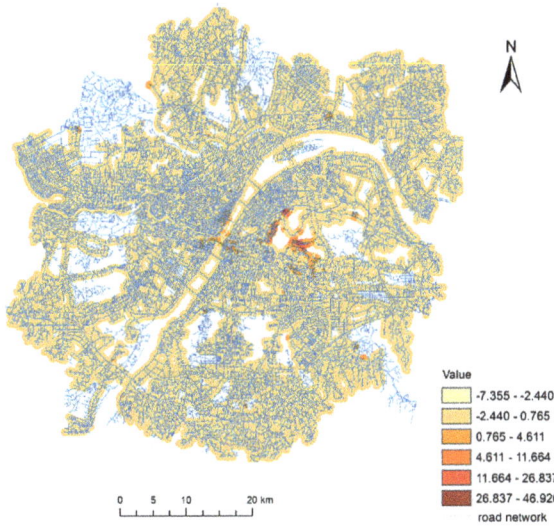

consistent with the scenic hotspots that the network kernel density is looking for, such as the East Lake Scenic Area.

9.3.4   Evaluation

Referring to the study of Luscher and Weibel (2013), the precision, recall, and F1-score are calculated, and the specific calculations are as follows (Eqs. 9.3, 9.4, 9.5).

$$precision = \frac{a_{overlap}}{a_{computed}} \qquad (9.3)$$

$$recall = \frac{a_{overlap}}{a_{comparative}} \qquad (9.4)$$

$$F_{1-score} = 2 * \frac{precision * recall}{precision + recall} \qquad (9.5)$$

Among them, $a_{computed}$ is the calculated area of some type of urban functional zone, $a_{comparative}$ is the area of the urban functional zone that is the reference, and $a_{overlap}$ is the area of overlapping parts between urban functional zone of a certain type researched and the functional area of the type referenced. The higher the F_1-score, the higher the coincidence of the two functional areas.

9.4   Experiment and Discussion

The proposed method has been applied to identify the urban functional zones of Wuhan in China. For the study area, the experiment datasets include POIs and road network data of Wuhan in 2014, which are both official data from Wuhan Land Resources and Planning Bureau.

9.4.1   Sensitivity Analysis of Core Parameters

Bandwidth is a significant parameter in the network kernel density estimation method. At present, there is no systematic method for determining the bandwidth in the network kernel density estimation. Based on experience, 300, 600, 900, and 1200 m are generally used as the bandwidth (Yu et al., 2015). Therefore, in this experiment, the network kernel density estimation and Kriging interpolation results are tried under various bandwidths, proving 600 m is the most appropriate. Figure 9.4 is a comparison of four different distance attenuation thresholds for the same area of an industrial POI (Guanggu Software Parks). It can be seen from Fig. 9.4 that the distance attenuation value of 300 m causes more high-value or low-value regions in the density distribution results, which is suitable for revealing the partial characteristics of the density distribution. But due to the excessively high value regions and fragmentary distribution, it does not adapt to the scale of the urban road network so this threshold is excluded from dividing the functional area. Although the distance attenuation thresholds of 900 and 1200 m can make the hotspots more apparent at the global scale, the density distribution results often exceed the scope of the urban road network and involve other types of POI high-value areas. Thus, these two thresholds are also eliminated. In addition, larger distance attenuation thresholds sometimes combine polycentric distributions with a single-centric distribution. In the case of a distance attenuation threshold of 600 m, the high-value areas in the density distribution results are relatively concentrated, and they do not classify some other areas into industrial type areas as in the 900 and 1200 m threshold cases, so a 600 m distance attenuation threshold is adopted. The distance attenuation value should be positively related to the discreteness of facilities. For sparse point facility distributions, larger distance attenuation values should be used, while for dense point facilities, we need to consider smaller distance attenuation values [20]. At the same time, the greater the distance attenuation threshold, the smoother the created density surface.

9.4.2   K-Means Results

According to the statistics of the POI data in the clustering area (Table 9.1), the landscape, commerce, housing, industry, and education categories account for the

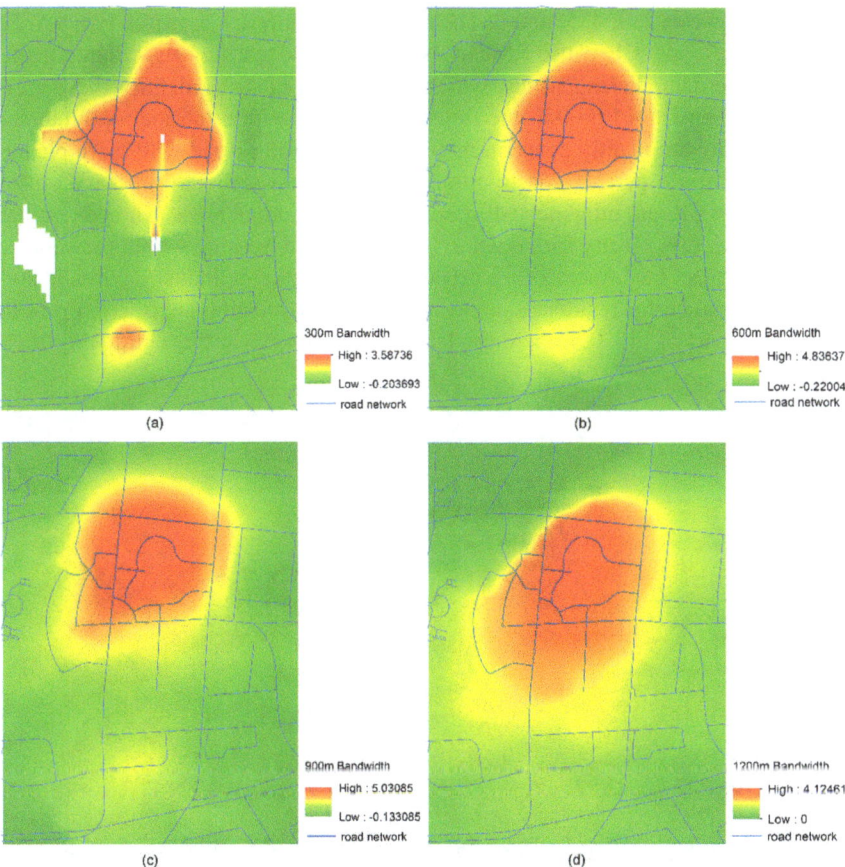

Fig. 9.4 Kriging interpolation results of network kernel density estimation for different bandwidth near Guanggu software parks

Table 9.1 The proportion of various types of POI in each cluster area

Category	Residence	Commerce	Education	Industry	Landscape
1	0.3289	0.4815	0.1790	0.0091	0.0014
2	0.0525	0.0447	0.0486	0.0039	0.8501
3	0.0506	0.8987	0.0414	0.0092	0
4	0.8555	0.0833	0.0579	0.0033	0
5	0.0116	0.0463	0.0386	0.9034	0
6	0.1046	0.0968	0.7606	0.0095	0.0285
7	0.2830	0.2453	0.1415	0	0.3301
8	0.4878	0.1360	0.3590	0.0135	0.0037

most in categories 2, 3, 4, 5, and 6. The clustering results are shown in Fig. 9.5. For categories 1, 7, and 8, because KMeans is unsupervised learning algorithm, affected easily by the data characteristics, which makes it difficult to identify these areas. As can be seen from Fig. 9.4, the functional scopes within the metropolitan area of Wuhan City are concentrated in the Third Ring Road with more planned policies and more perfect construction. The industrial areas are primary outside the Third Ring Road, interspersed by scattered commercial areas. The government stipulates that it is strictly forbidden to build complete housing within the scope of industrial project land, therefore, a certain range around the industrial area is not a reasonable choice for residential areas, science and education areas, etc. From the lower right part of Fig. 9.5, there are large industrial sites at the Third Ring Road,

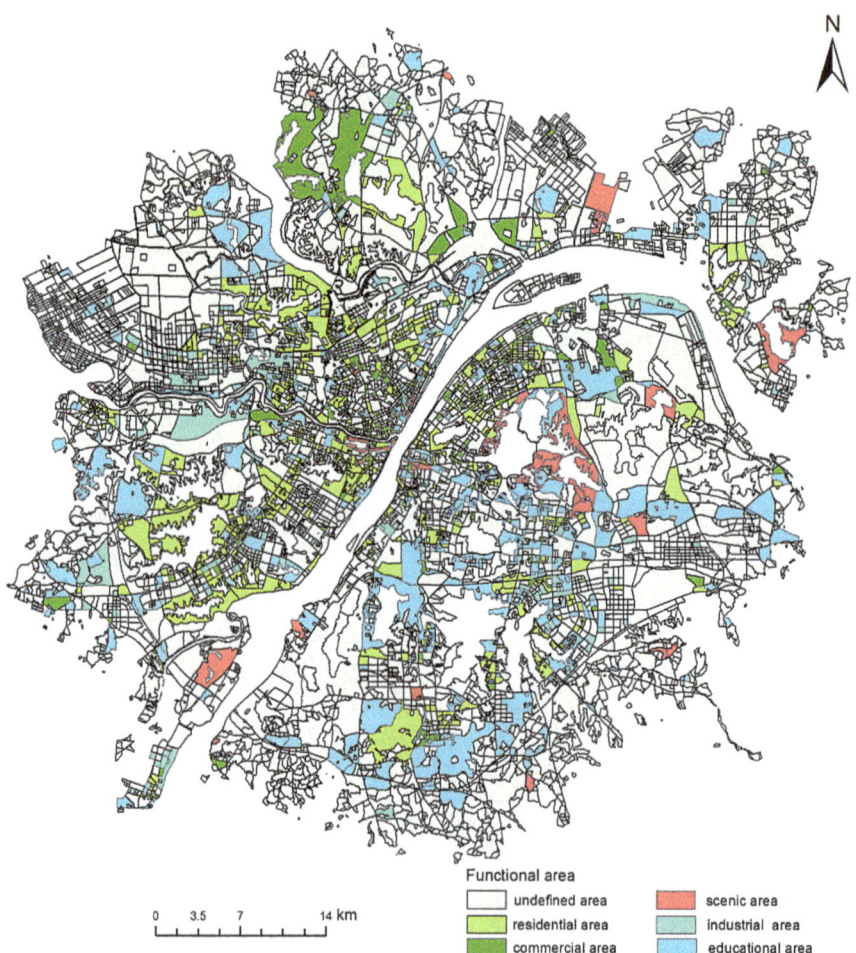

Fig. 9.5 The division results of the urban functional zones based on the POI proportion

but a small piece of science diocese is located not far north. Research, education, scenic spots and residential areas are clearly distributed in space, consistent with the current situation of urban land use.

However, as shown in Fig. 9.6, it has challenges to accurately identify most of the areas in the study, especially in education-residence, landscape-residence and commercial-residential areas. Therefore, kernel density estimation method is also wanted. Identifying the density values of various types of POIs in urban space facilitates the accurate identification of various functional areas in the city's core areas and agglomeration areas according to the distribution characteristics of urban road network grades and POI.

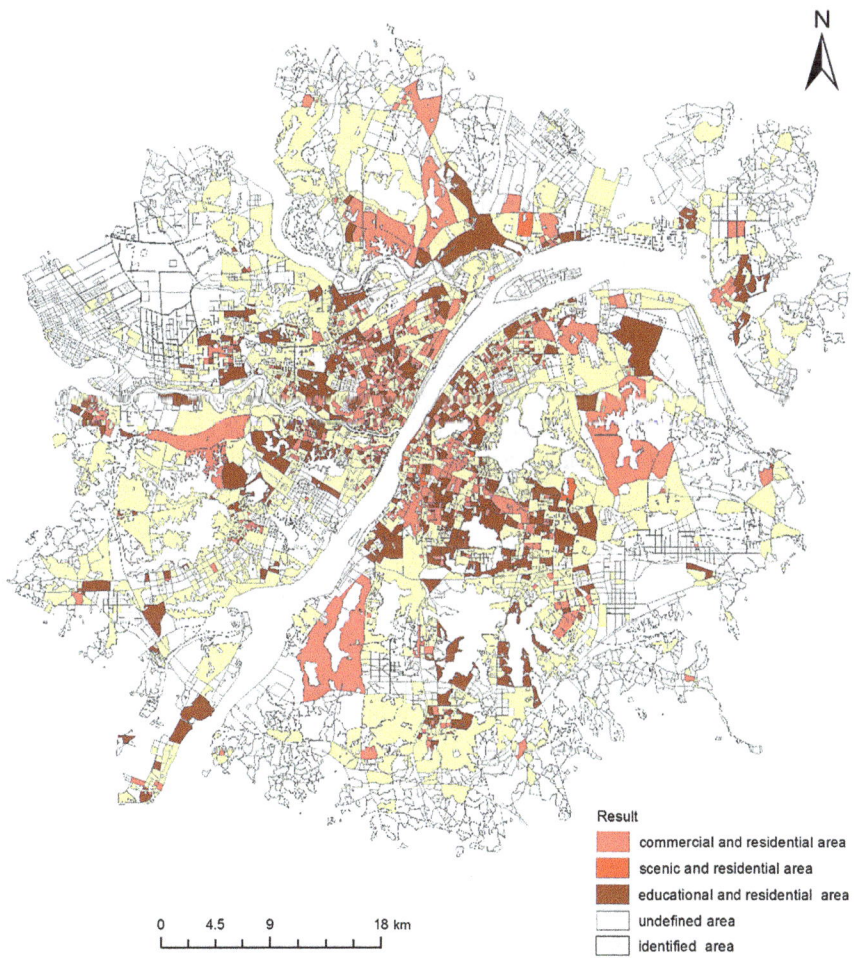

Fig. 9.6 Areas that cannot be identified after clustering

9.4.3   Findings of Interpolated High-Intensity Density Area

Because the K-Means results cannot accurately identify the functional characteristics of mixed region with POI types, the estimated density of the various types of POI data along the road network provides the foundation to estimate the kernel density, which is advantageous in finding the high core intensity regions of various types of POI. The network kernel density is estimated for each POI representing the various functional areas. Each type of POI uses multiple distance attenuation thresholds for multiple experiments. The distance attenuation threshold parameters for the same type of POI data are set to 300 m, 600 m, 900 m and 1200 m respectively. Then, according to the different distance attenuation thresholds used in the previous kernel density estimation, the corresponding search radius is determined. Finally, the functional area is divided by using the interpolation results of five types of POIs and the graded urban road network, respectively. The interpolation results of a certain type of POI is used to define the main distribution location and range, and the area with the higher interpolation result value is the centralized distribution place of this type of POI.

   With the gradual improvement of urban infrastructure, and education is closely related to living and business. Because the number of POIs, such as school, on the map is far less than that of business or residence, the area of the schools is often very large. This has led to the difficulty of accurately identifying commercial, residential and educational areas in some places. To further identify the business, education, and living in urban distribution, we need to conduct deep-level mining. With the help of a network kernel density analysis and Kriging interpolation method, shown in Figs. 9.7, 9.8 and 9.9. From Fig. 9.6, we can derive that, the regions which are unable to be accurately identified in the clustering results can be accomplished well in Figs. 9.7, 9.8 and 9.9 as high-density regions for urban education, residence, and business.

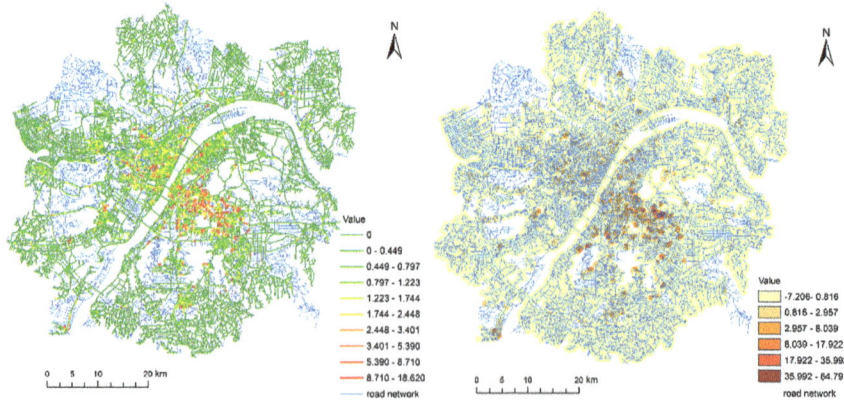

Fig. 9.7   Network kernel density and kriging interpolation results of education & research

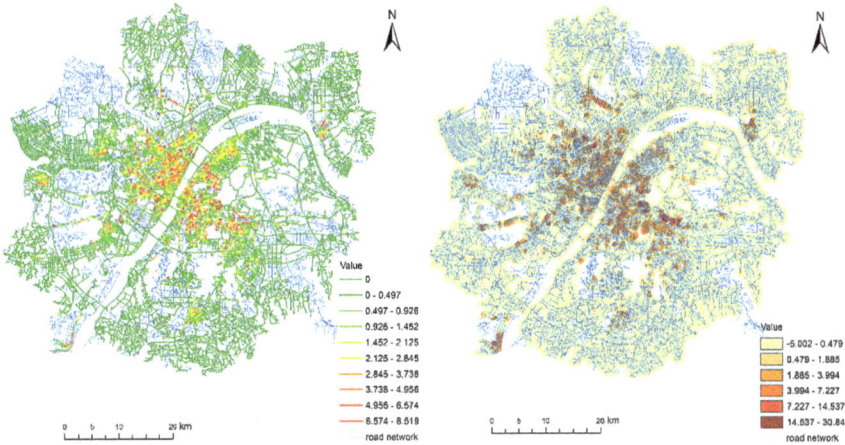

Fig. 9.8 Network kernel density and kriging interpolation results of residence

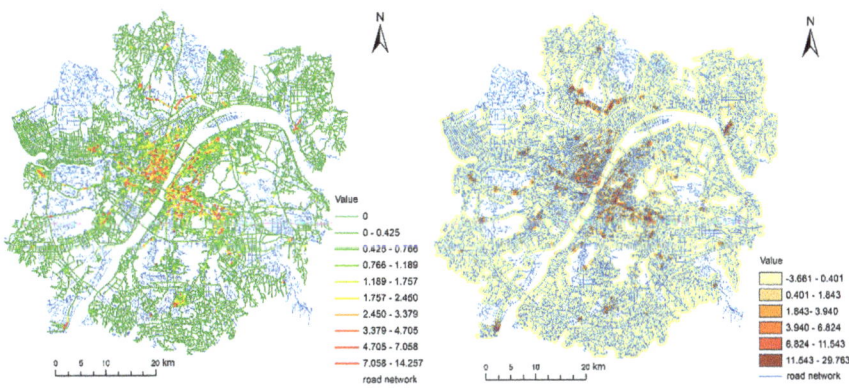

Fig. 9.9. Network kernel density and kriging interpolation results of business

The roads in the urban road network are classified into different levels according to the "City Road Engineering Design Code" (CJJ37-2012), and the graded urban road network is used to accurately define the precise boundaries of the functional areas. The boundaries of the functional areas are mainly confirmed by the main road or sub-road division. Yuan has summarized three methods for segmenting urban areas (Yuan et al., 2015). One is grid segmentation, the second is segmenting in terms of importance based on graded urban road network, and the last is segmentation applying morphological methods. Because both POIs and urban functional zones are closely linked with the urban road network, the approach of a hierarchical urban road network to segment urban areas is suitable to determine the functional area's boundary.

9.4.4 Evaluation

In terms of the quantitative evaluation method in Sect. 9.3.4, the results of functional area recognition are quantitatively evaluated. The vector data used as the reference functional area's range is the land use status data in the urban development circle of Wuhan City. Based on the requirements of data confidentiality, the scope of the reference functional area determined based on the data cannot be directly displayed in this paper. The results of the quantitative evaluation on the divided functional areas are presented in tabular form, as shown in Table 9.2.

As can be seen from Table 9.3, commercial areas, industrial areas, science education areas, and scenic areas, other than residential areas, have certain accuracy rates. The reason for the low accuracy of residential areas may be that the distribution of residential POI data is relatively even and scattered, resulting in a smaller difference in the Kriging interpolation results and a smaller peak. Therefore, both the

Table 9.2 Qualitative Evaluation

Type of functional area	Result diagram	Matching diagram	Regional subject function
Commercial area			Hanzheng street, Wuguang, Jianghan road business circle
Residential area			Triumphant, Jianghong garden, Nanhu garden and other residential districts
Industrial area			Xinlong industrial park, Dongfang industrial park and other industrial sites
Science education area			Wuhan University, Wuhan University of Technology, Central China normal university and other institutions of higher learning
Scenic area			Shouyi cultural area

Table 9.3 Quantitative evaluation

Functional area of city	Precision/%	Recall/%	F1-score
Commercial area	32.73	96.66	0.49
Residential area	44.72	26.72	0.33
Industrial area	86.7	49.46	0.63
Science education area	42.09	76.2	0.54
Scenic Area	93.46	91.31	0.92

number and area of the divided areas are significantly lower than the Wuhan residential area during the development of the land use in the circle. At the same time, the K Means clustering results indicate that the main reason why the type of functional areas cannot be accurately identified in Fig. 9.6 is that education-residential, commercial-residential relationships are closely related, resulting in the dispersion of urban living POIs and the inability to identify these clustered regions accurately. It proves the dispersal of residential areas.

9.5 Conclusion

This paper proposes a method based on POI feature and network kernel density. The research area is divided into 8 types of regions through the K-Means, the urban road network data and POI data. Then we use the network core density and The Kriging interpolation method to further identify urban functional zones. The experimental results show that: (1) The urban areas divided by fuzzy morphological operators can be used to identify functional areas such as landscapes, industries, etc., and bring into correspondence the current status of urban land use. (2) Due to the sensitivity of K-Means, the results in the incomprehensible identification of mixed regions, network kernel density and Kriging interpolation can not only effectively separates core functional areas such as education, residence, and business, but also demonstrates that education, residence, and business have significant relevance. (3) The functional zones of education, residence, business, landscape, and industry can be effectively determined and identified by the K-Means, network kernel density, and Kriging interpolation methods.

Experiments on Wuhan City show that this method is wieldy and accurate. Further research is planned to improve the existing approach to overcome the disadvantages of failing to divide all urban areas into specific functional areas and to recognize mixed areas.

References

Chi, J., Jiao, L. M., Dong, T., et al. (2016). Quantitative identification and visualization of urban functional area based on POI data. *Journal of Geomatics, 41*(2), 68–73. (in Chinese).

Gao, S., Janowicz, K., & Couclelis, H. (2017). Extracting urban functional regions from points of interest and human activities on location-based social networks. *Transactions in GIS, 21*, 446–467.

Jiang, G., Hu, F., & Shi, L. (2016). Urban functional area identification based on call detail record data. *Journal of Computational and Applied, 36*(7), 2046–2050. (in Chinese).

Jing, Y. (2008). Remote sensing semantic model for city planning. *Computer Applications, S1*, 348–435. (in Chinese).

Kang, Y. H., Wang, Y. Y., & Xia, Z. J. (2017). Identification and classification of Wuhan urban districts based on POI. *Journal of Geomatics, 43*(1), 1–5. (in Chinese).

Li, C. H., Sun, Z. H., Chen, G., et al. (2004). Kernel density estimation and it's aplication to clustering algorithm construction. *Journal of Computer Research and Development, 41*(10), 1712–1718. (in Chinese).

Luscher, P., & Weibel, R. (2013). Exploiting empirical knowledge for automatic delineation of city centres from large-scale topographic databases. *Computers, Environment and Urban Systems, 37*, 18–34.

Shili, C., Haiyan, T., Xuliang, Li., et al. (2016). Discovering urban functional regions using latent semantic information: Spatiotemporal data mining of floating cars GPS data of Guangzhou. *Acta Geographica Sinica, 71*(3), 471–483. (in Chinese).

Song, X. T., Pu, Y. X., Liu, D. W., et al. (2015). Mining urban regional functional attributes by pedestrian trajectory. *Acta Geodaetica Et Cartographica Sinica, 44*(S1), 82–88. (in Chinese).

Steiniger, S., Lange, T., Burghardt, D., et al. (2008). An approach for the classification of urban building structures based on discriminant analysis techniques. *Transactions in GIS, 12*(1), 31–59.

Sun, R. H., Lv, Y. H., Chen, L. D., et al. (2013). Assessing the stability of annual temperatures for different urban functional zones. *Building and Environment, 65*, 90–98.

Wenhao, Y., & Tinghua, A. (2015). The visualization and analysis of POI features under network space supported by kernel density estimation. *Acta Geodaetica Et Cartographica Sinica, 44*(1), 82–90. (in Chinese).

Xiang, Y. (2014). *Discovering zones of different functions using bus smart card data and points of interest: A case study of Beijing.* Zhejiang University.

Yan, W., Zhenbai, S., & Peilin, W. (2009). A study on spatial clustering of urban function partition. *Areal Research and Development, 28*(1), 27–31. (in Chinese).

Yu, W. H., Ai, T. H., Shao, S. W., et al. (2015). The analysis and delimitation of central business district using network kernel density estimation. *Journal of Transport Geography, 45*, 32–47.

Yuan, N. J., Zheng, Y., Xie, X., et al. (2015). Discovering urban functional zones using latent activity trajectories. *IEEE Transactions on Knowledge and Engineering, 27*(3), 712–725.

Zhi, D. (2010). *Spatial clustering algorithm on urban function oriented zone.* Sichuan Normal University.

Chapter 10
Evaluation Methodology on Industry-City Integration Degree of China National High-Tech Industrial Development Zones: A Case Study of Hubei Province

Pei Chen and Yaping Huang

10.1 Introduction

Along of reform and opening up policy, the construction of high-tech zones boom, sparked a decade of lively discussion on the issue of industry-city integration. Commonly, high-tech zones (or industrial park) were located at urban fringe area. The relatively independent high-tech zone has led to the division of industrial development and urban function, resulting in varying degrees of industry-city separation. The main characterization of industry-city separation is that the development of living space lags behind the development of production space, which implicates the construction of urban function. Industry is the basis of urban development, while the city is the carrier of industrial development (Daogang, 2011). It will be an empty city without industry; equally it will be a lifeless collection of machines without a city (Yishao, 2016).

In recent years, industry-city integration is gradually becoming a new development concept of high-tech zones, high-tech zones began more likely to be a comprehensive urban area than a single industrial base. Over time, it has been a magic weapon of transformation for high-tech zones to promote the interactive development of industrialization and urbanization (Du, 2014). Nevertheless, industry-city separation has spawned numerous problems on many aspects, such as long-distance commuting, expensive occupancy costs, imperfect supporting facilities, lack of living

P. Chen (✉) · Y. Huang
School of Architecture and Urban Planning, Huazhong University of Science and Technology, No. 1037, Luoyu Road, Hongshan District, Wuhan, Hubei, China
e-mail: chenpeihust@hust.edu.cn; d201577764@hust.edu.cn

Y. Huang
e-mail: hust_hyp@sina.com

© Springer Nature Switzerland AG 2021
W. Li et al. (eds.), *Human-Centered Urban Planning and Design in China: Volume I*,
GeoJournal Library 129, https://doi.org/10.1007/978-3-030-83856-0_10

entertainment resources, poor air quality and sound environment, which can hardly get targeted solutions merely from the concept of industry-city integration.

The discussion on industry-city integration started from the relationship between industrialization and urbanization. Industry-city integration is broadly interpreted as the integration of industrialization and urbanization, and narrowly interpreted as the integration on the space of industry and city. China's research on industry-city integration has been mainly from these four aspects, the concept, characteristics and mechanisms, evaluation method, and strategies.

The study on the concept of industry-city integration shows a gradually deepening process, specifically from narrow and static to broad and dynamic (Fu, 2015). On the other hand, the classification study has been lacking of enough attention or in-depth exploration. The contents of industry-city integration include aspects on economic, social, cultural, and spatial, etc.; the spatial types can be roughly divided into 4 main types, central area neighboring type, fringe area growing type, new district adjoining type, and independent type (Gangbiao, 2011). Discussions on the essence of the integration differ. There is a point of view that it's about the integration of living and employment, which reflects on the spatial relation between the urban communities and industrial parks (Hua, 2011). In recent years, however, many researchers believe that industry-city integration shows a state of scientific development, which blend intensive and efficient industrial space and livable living space, with ecological space (Wenbin et al., 2012; Wang et al., 2014; Du, 2014). Nowadays, as a consensus, the understanding on industry and city in the concept of industry-city integration is much more thorough, as the industry represents function of production, meanwhile the city represents function of service (including services on living and manufacture). The spatial level of discussion turns from the specific area in individual city even to the regional range (Du, 2014). The meaning of integration is interpreted as a comprehensive and chronic process of interaction, evolves into a complex or composite form containing multiple variants through cross infiltration (Huadong, 2012).

The study on characteristics and mechanisms of industry-city integration roughly remain consistent without highlighting the uniqueness. Previously, some researchers believe that the space medium during integration process is the choice of leading industry, the convergence to master plan, locational function, spatial structure and land use, facilities and transportation organizations, etc., which leads to a formation of regional innovation system, based on the interaction of industry, city, residents, and enterprises, relying on the elements, such as land and traffic, etc. (Gangbiao, 2011). Later, researchers tend to summarize the characteristics into 3 aspects: the functional match ability, the spatial miscibility, and low carbon transportation (Wei Jinlan, etc., 2014). About the mechanism, as a commonly cognition, is divided into internal and external power. The former includes industrial association, technological innovation, etc (Tang, 2014). The latter includes planning guidance, policy support, legal supervision, etc. (Yongming et al., 2013).

Although the research on evaluation and measurement of the industry-city integration has been widely concerned in recent years, most of them were not comprehensive enough, usually focused on the coordination relationship between industrialization and urbanization (Table 10.1). Among them, there're two evaluation methods with more reference value. One of them applied the entropy evaluation method by

Table 10.1   The evaluation index systems of industry-city integration

Time	Researcher	Method	The evaluation index system
2007	Lin Gaobang	Granger causality validation, regression analysis	(a) Construction output value represents urbanization, mechanization output value represents industrialization (b) Take a granger causality validation between construction output value with Per Capita GDP, etc. (c) Take a granger causality validation between mechanization output value with industry employment proportion, etc.
2007	Zhao Yan	System Dynamics (SD)	(a) Index system consists of industrialization index system and urbanization index system (b) Industrialization index system has 4 categories, development structure, economic benefit, etc.; 21 indexes, per capita GDP, industrial added output value, etc. (c) Urbanization index system has 4 categories, social, population, etc.; 27 indexes, per capita GDP, service output value ratio, etc.
2011	Gao Gangbiao	Analytic Hierarchy Process (AHP), Delphi technique	Index system has 7 categories, master plan, policy, GDP, etc.; 26 indexes, service employment ratio, road area per capita, etc.
2013	Su Lin et al.	Analytic Hierarchy Process (AHP), Fuzzy Evaluation Method (FEM)	Index system has 15 categories, human resources, infrastructure, environmental protection, etc., 73 indexes
2013	Wang Xia et al.	Factor analysis, Entropy evaluation method	(a) Turn the original 77 indexes into 13 factors by factor analysis (b) Apply the entropy evaluation method for weighted assignment (c) 13 factors, economic scale and efficiency, culture and activity, etc.
2014	Wang Fei et al.	Analytic Hierarchy Process (AHP), Factor analysis	Index system is made up of 2 categories, industrial degree and urbanization degree; 12 indexes, per capita GDP, household area ratio, etc.
2014	Tang Xiaohong	Grey Relational Analysis (GRA)	Index system is made up of 38 indexes, industry development degree, medical facility satisfaction degree, employment density, etc.
2014	Jiang Yining	Analytic Hierarchy Process (AHP)	Index system is made up of 3 categories, industrial integration, special integration and functional integration; 53 indexes, road area ratio, Engel coefficient, etc.
2014	Li Guanghui	Analytic Hierarchy Process (AHP)	Index system has 4 categories, industrial development, city construction, humanization development and interactive degree; 38 indexes, per capita GDP, population density, etc.
2015	Li Zhenshan	Analytic Hierarchy Process (AHP)	Index system has 4 categories, policy conformance, economic development, special mixture, environmental protection; 26 indexes, road area ratio, service employment ratio, etc.

(continued)

Table 10.1 (continued)

Time	Researcher	Method	The evaluation index system
2015	Fu Jing	Analytic Hierarchy Process (AHP), Fuzzy Evaluation Method (FEM)	(a) Index system consists of industrialization index system and urbanization index system (b) Industrialization index system has 4 categories, industrial structure, science progress, etc.; 25indexes, per capita GDP, infrastructure, etc. (c) Urbanization index system has 5 categories, living standard, employment, etc.; 39 indexes, industrial waste gas emission per capita, urbanization rate, etc.

Contents come from related references

extracting the common factor from the factor analysis, based on 56 China national high-tech industrial development zones (Wang et al., 2014). The other one, additionally concerned about the interactive relationship between people and industry, people and city, and added evaluation dimensions about humanization development and artificial city development (Li, 2014). The index system created by Li Guanghui, which applied the Delphi Technique, would likely to face the problems of factor superposition and assignment subjection. However, its consideration of interactive development is worthy of reference and deepening.

The study method on the strategy of industry-city integration gradually turns more dynamic and innovative, with the process of traditional industrial parks transform to more comprehensive ones, both in structure and function. Some researchers take Jinan High-tech Zone as an example, put forward specific strategies, such as structural optimization, function upgrading, industry promoting, traffic organization, facilities supporting, urban style construction and low-carbon development (Liu et al., 2012). Some researchers propose strategies as functional compound, facilities comprehension, layout mixture, etc. (Chang et al., 2012). Some researchers propose 3 steps of strategies, industrial advancing strategy, social service strengthening strategy, organic growth strategy (Rongzeng, et al., 2013). Other researchers put forward 9 planning strategies, making position, functional compound, space stitching, industrial aggregation, planning coordination, structural coupling, humanistic integration, facility reconciliation and land use mixture (Dong et al., 2014). In recent years, more and more researchers turn focus on compromise urban districts or technological cities. Some researchers discuss about the development model and mechanism of the development of technological cities, and analyze the contents of spatial, industrial and ecology evolution, management mode, etc. Yang Chang explores the transformation and upgrading of technological cities, based on the key problems, such as secondary development of land use and feature creation, etc. (Chang, 2014). Zhu Yun, take Suzhou high-tech zone for an example, proposes regional cultural diversity, ecological penetration, multi-level services and other strategies, aiming at multi-dimensional goals, industry, innovation, culture, ecology, and social (Zhu, 2015).

Fortunately, the previous misunderstandings are facing discrimination and reflection, but which still need to worry about are garble and blindness. According to the research review above, which on the four aspects, the evaluation methods are numerous, but most of them have limited contribution, while the strategies are lack of pertinence and operability because of the absence of analysis on the results of evaluation. Therefore, the evaluation and strategies of industry-city integration need more in-depth thinking and research. In addition, the process between evaluation and strategies should be attached more importance on.

10.2 Methodology

10.2.1 Research Samples and Data Sources

Vast territory and regional disparity, this research is appropriate to a regional perspective. As the midland of China, Hubei Province plays a linking role both in geographical location and developmental sequence. Wuhan, provincial capital of Hubei, is the regional hub city of China. Wuhan East Lake Self-dependent Innovation Demonstration Area is the second one after Beijing Zhongguancun. Therefore, Hubei case offers reference value to the vast mid-west region of China, especially during the national-wide transformation era.

In this research, seven national high-tech zones in Hubei Province were selected as the research samples, namely: Wuhan East Lake Self-dependent Innovation Demonstration Area, Yichang High-Tech Industrial Development Zone, Xiangyang High-Tech Industrial Development Zone, Xiaogan High-Tech Industrial Development Zone, Suizhou High-Tech Industrial Development Zone, Jingmen High-Tech Industrial Development Zone and Xiantao High-Tech Industrial Development Zone.[1]

The data come from the statistics yearbooks of each city, statistics annual bulletin and related governments' official websites of 2015.

10.2.2 Research Methods

Based on the references, this research refers to the evaluation system created by Wang Xia, et al., and combines with the perspective of humanization and interaction (Li, 2014). The original index system is composed of three levels, 3 first-grade indicators (industrialization, urbanization and industry-city separation), 18 s-grade indicators and 77 third-grade indicators (Wang et al., 2014). Total indicators are reduced from 77 to 38, after data validation, variable correction and standardization. Through factor

---

[1] Xianning high-tech zone and Huanggang high-tech zone were formally approved as national level by State Council of PRC in Feb., 2017. The number of national high-tech industrial development zones in Hubei province has been 9 since then. But the latest annual data at the time of research is the year of 2015; this research is based on the data of 2015. And the samples are the 7 national high-tech industrial development zones in 2015.

analysis of the 38 indicators of 56 high-tech industrial development zones during 2009–2012, the final 13 factors and their corresponding weights are determined. 13 factors are divided into 3 categories according to the character of them: hardware foundation factors (technology & innovation, economic scale and efficiency, social public utilities, and infrastructure); software environment factors (culture, education and health, culture and activities, staff education structure, and environmental protection); interactive factors (output of science and technology activities & innovative service, liabilities and assets, industrialization & social urbanization, capitalization & population urbanization, industrialization & population urbanization) (Table 10.2).

Firstly, carry out every factor and summarize them, rank the integration degree of all samples. Then, respectively rank the hardware foundation index, software environment index and interactive index of all samples. Finally, successively make the comparison, clustering, and analysis on the four results above.

Each value of factor should be standardized according to the following method. If the value is an aggregate indicator, above all, do a per capita processing. For positive factors, the standard value equals to actual value/target reference value, when the standard value is above or equals to 1, make standard value equals to 1. For negative factors, the standard value equals to target reference value/actual value, when the standard value is above or equals to 1, make standard value equals to 1. For neutral factors, apply secondary standardization method. Among them, the target reference value is determined as follows: choose the maximum if it's positive

Table 10.2 Evaluation index system of industry-city integration of China national high-tech industrial development zones

Composite factors	Category of factors	Number	Name of factors	Character	Weights (Wj) (%)
	Hardware foundation factors	FAC 1	Technology & innovation	Positive	4.3
		FAC 2	Economic scale and efficiency	Positive	6.3
		FAC 3	Social public utilities	Positive	10.1
		FAC 4	Infrastructure	Positive	11.3
	Software environment factors	FAC 5	Culture, education and health	Positive	6.3
		FAC 6	Culture and activities	Positive	5.5
		FAC 7	Staff education structure	Positive	10.8
		FAC 8	Environmental protection	Negative	7.6
	Interactive factors	FAC 9	Output of science and technology activities & innovative service	Positive	13.6
		FAC 10	Liabilities and assets	Negative	8.4
		FAC 11	Industrialization & social urbanization	Neutral	7.4
		FAC 12	Capitalization & population urbanization	Positive	4.1
		FAC 13	Industrialization & population urbanization	Negative	4.2

(individual disperse value can be ignored), the minimum if it's negative, and the average of the same period if it's neutral.

10.3 Results

10.3.1 Composite and Category Results

Calculated integration index and the individual value of all samples, converted into percentages are as shown in Table 10.3.[2]

The rank of integration shows the structural feature of three echelons (Fig. 10.1). In descending order of scoring, the first echelon includes Wuhan and Yichang; the second echelon includes Xiaogan, Suizhou, and Xiangyang; the third echelon includes Jingmen and Xiantao. Specifically, integration degree of Wuhan or Yichang is above 75, the degree of second echelon is in the interval of 60–70, the degree of third echelon is in the interval of 50–60. The average score difference in each echelons are sequentially 6.47, 1.49, and 1.79. The average score difference between every two adjacent echelon are sequentially 15.52, and 9.91.

Similarly, the hardware foundation index, software environment index and interactive index also show the structural features of three echelons (Figs. 10.2, 10.3 and 10.4). The composite and each category results show some visible characteristics, which can be concluded to several integration patterns (Fig. 10.5).

10.3.2 Integration Patterns

Echelon structure and data interval of composite index represents the degree of industry-city integration. In particular, the first, second and third echelon of composite index serially represents the degree of integration, half-integration, and not-integration. The relationship among composite index, hardware foundation index, software environment index and interactive index, represents the status of industry-city integration. In particular, there are types of equilibrium, fluctuation, and disequilibrium.

Considering degree and status of industry-city integration simultaneously, 7 samples are classified into 5 patterns: the pattern of equilibrium & integration (Wuhan), the pattern of equilibrium & half-integration (Xiaogan), the pattern of fluctuation & integration (Yichang), the pattern of fluctuation & not-integration (Jingmen and Xiantao), and the pattern of disequilibrium & half-integration (Xiangyang and Suizhou) (Fig. 10.6).

---

[2] In order to simplify the form of expression, the following by the city name instead of the city's high-tech zone name.

Table 10.3 Evaluation results of industry-city integration of China national high-tech industrial development zones in Hubei

Name of factors	Wuhan	Yichang	Xiaogan	Xiangyang	Suizhou	Jingmen	Xiantao
Technology & innovation	4.30	0.99	4.30	2.06	0.47	0.82	0.73
Output of science and technology activities & innovative service	13.60	12.78	11.15	13.60	11.83	9.93	5.98
Economic scale and efficiency	6.30	6.30	2.33	3.34	6.30	4.47	3.91
Social public utilities	2.42	9.09	10.10	8.18	10.10	7.58	5.45
Culture, education and health	6.30	3.72	3.15	4.54	3.78	4.16	3.53
Liabilities and assets	6.05	5.21	6.89	6.05	6.22	7.22	8.40
Culture and activities	4.79	5.50	0.33	4.24	0.17	0.33	0.17
Staff education structure	8.10	7.88	7.34	10.80	3.24	6.26	4.54
Infrastructure	11.30	11.30	2.26	3.84	4.29	0.11	1.24
Environmental protection	7.60	4.64	5.62	3.27	4.26	3.12	7.60
Industrialization & social urbanization	7.40	4.44	6.07	1.33	6.44	5.40	6.88
Capitalization & population urbanization	0.57	0.41	4.10	0.78	1.39	3.24	4.10
Industrialization & population urbanization	4.20	4.20	2.02	0.80	4.20	2.52	0.84
Integration index	82.93	76.46	65.66	62.82	62.69	55.16	53.37

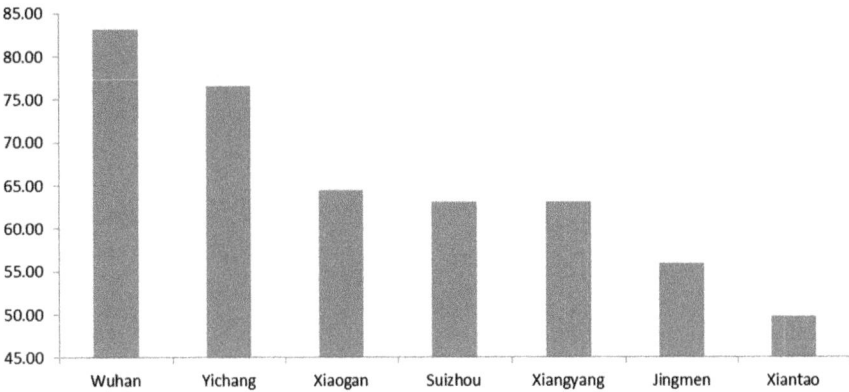

Fig. 10.1 Ranking of integration index

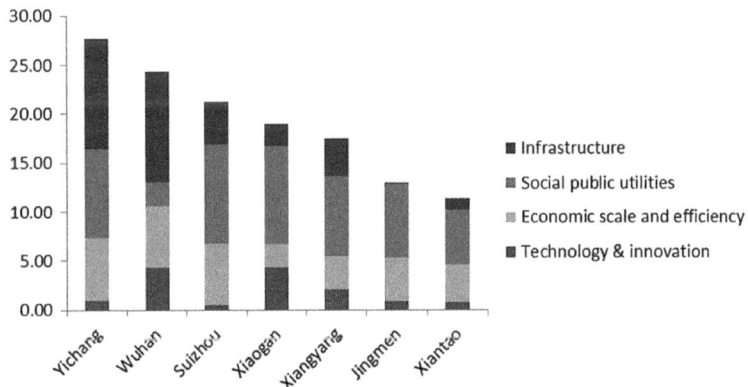

Fig. 10.2 Ranking of hardware foundation index

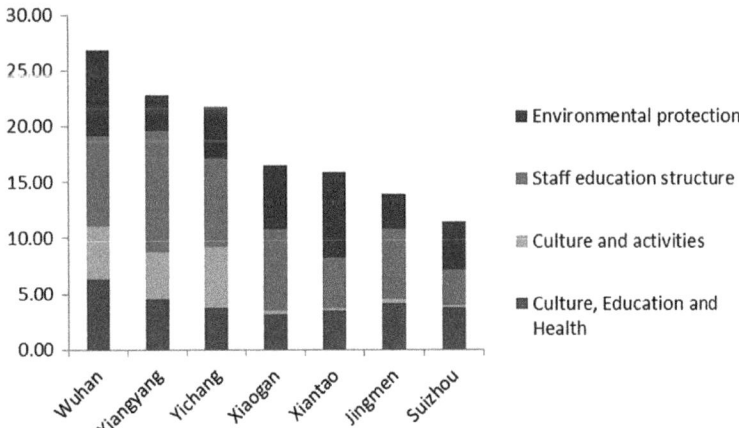

Fig. 10.3 Ranking of software environment index of integration evaluation

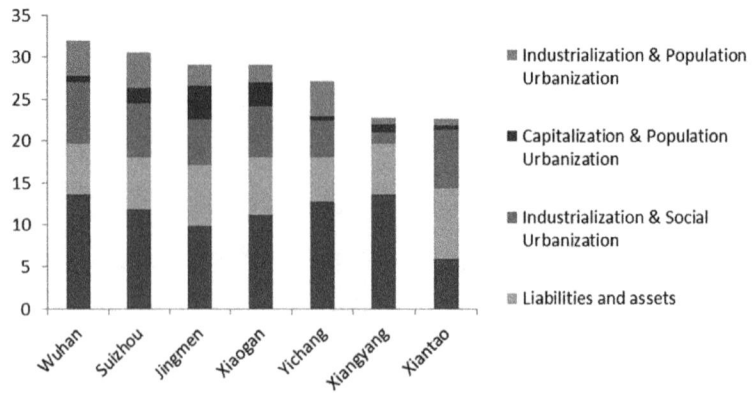

Fig. 10.4 Ranking of interactive index of integration evaluation

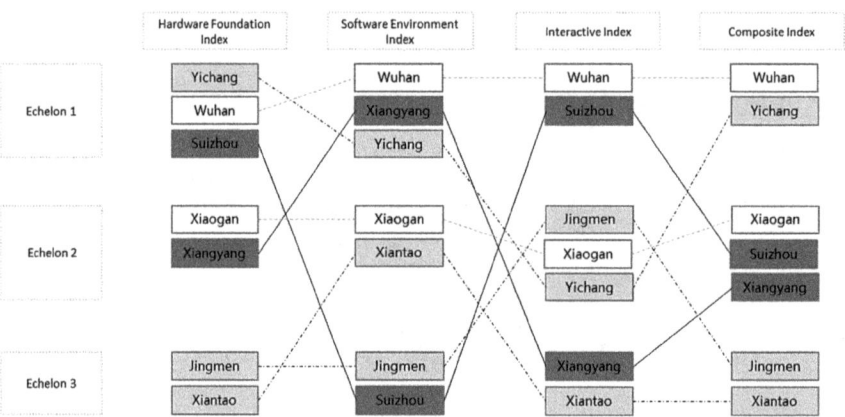

Fig. 10.5 Composite and category results of integration evaluation

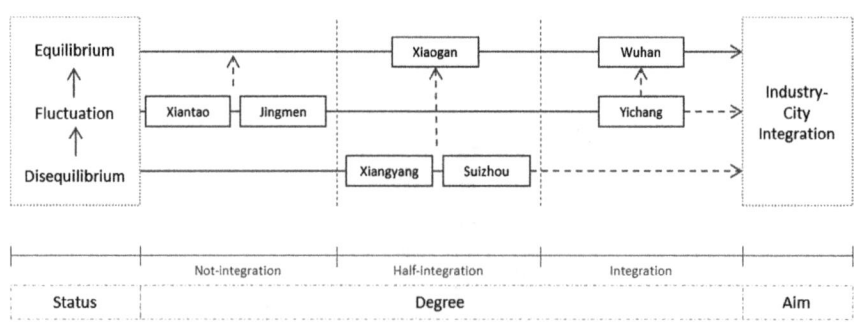

Fig. 10.6 Patterns of industry-city integration

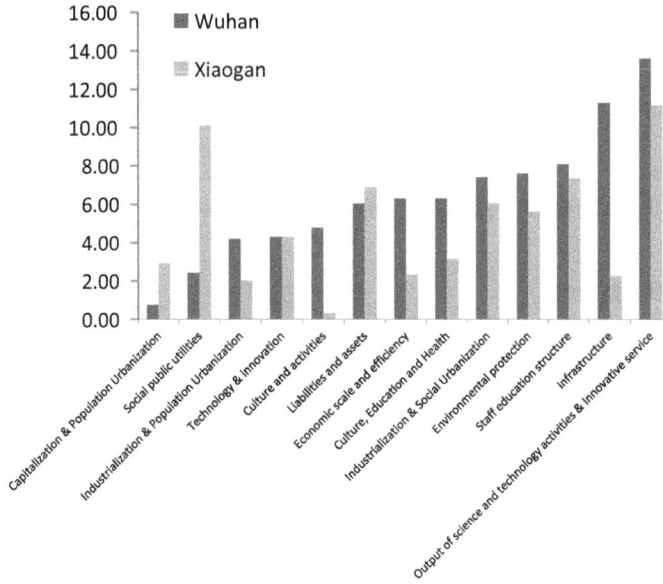

Fig. 10.7 Comparison between Wuhan and Xiaogan in individual factors

The same as the pattern of equilibrium, there is visually balanced gap between Xiaogan and Wuhan (Fig. 10.7). No matter the pattern of fluctuation & integration or fluctuation & not-integration, specifically, Yichang, Jingmen, and Xiantao, consolidated results tend to be balanced (Fig. 10.8). No matter Xiangyang, which three partly index are distributed in totally different echelons, or Suizhou, which three partly index are distributed in the first and third echelons, their consolidated results tend to be balanced, too (Fig. 10.9). Thus it can be seen; all patterns of integration tend to develop as the pattern of equilibrium, in order to upgrade the degree of integration.

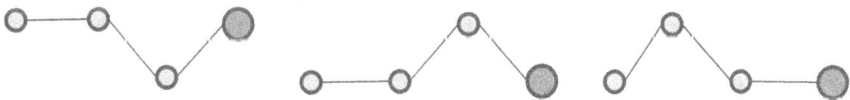

Fig. 10.8 The patterns of fluctuation, Yichang, Jingmen, and Xiantao[3]

---

[3] Figure 10.8 is simplified from Fig. 10.5. The three small circles refer to the level of results on hardware foundation, software environment and interaction; and the big circle refers to the composite result. As the visually obvious feature, only one among four results makes a one-level difference, nearly to be balanced.

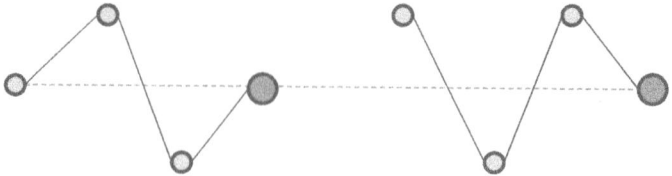

Fig. 10.9 The patterns of disequilibrium, Xiangyang and Suizhou[4]

10.4 Discussion

This research picks out dominance elements or potential elements of not achieving integration through both crosswise and vertical orientation analysis. Then, it proposes targeted strategies.

10.4.1 Accelerating Optimization, Potential Drawbacks Elimination

Yichang is supposed to follow the strategy of accelerating optimization to eliminate its potential drawbacks.

As the pattern of fluctuation & integration, the industry-city integration of Yichang seems qualified at first glance. However, there is still potential drawback when searching for the root cause of fluctuation in integration. The potential drawback is capitalization & population urbanization. Although it's unlikely to affect the integration process overall, it exactly is the essential element that pulls Yichang a gap with optimal sample, Wuhan. Additionally, the integration of Yichang is closely mature in degree, accelerating optimization is necessary, before the drawback turns solidified or normalized.

Capitalization & population urbanization represents the relationship between professionalization process of industry and population growth in surrounding districts. Considering the comprehensive development of both hardware and software equipment, the satisfaction of completed of facilities and environment is short of expectations. Key solution lies in humanistic orientation of policy making and planning. On one hand, Yichang should emphasize on a wider range of public participation, make better aware of real and urgent demand of different stakeholders through sufficient communication. On the other hand, apply dynamic evaluation and feedback mechanism to implementation, so as to follow up changes in the structure and demand of stakeholders. Thus, fundamentally recover the existing or even forecasting

---

[4] Figure 10.9 is simplified from Fig. 10.5. The three small circles refer to the level of results on hardware foundation, software environment and interaction; and the big circle refers to the composite result. As the visually obvious feature, the four results distribute in three levels, far from balance.

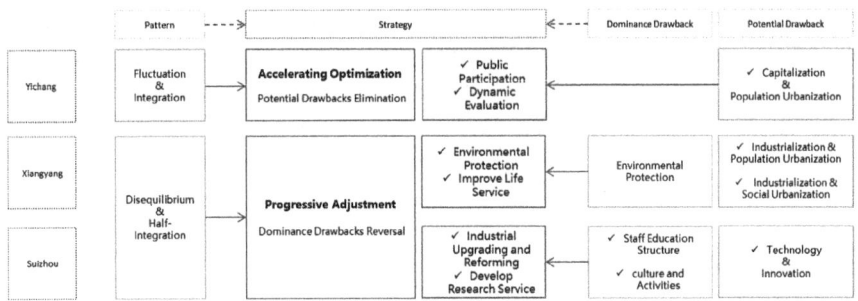

Fig. 10.10 Analysis process of strategies of Yichang, Xiangyang, and Suizhou

contradiction between supply and demand; upgrade the satisfaction of interaction environment in the integration process (Fig. 10.10).

10.4.2 Progressive Adjustment, Dominance Drawbacks Reversal

Xiangyang and Suizhou are supposed to follow the strategy of progressive adjustment to reverse their dominance drawbacks.

Xiangyang's comparative weakness, the interactive aspect, restricts the development of its industry-city integration. Among which industrialization & population urbanization and industrialization & social urbanization are the dominance drawbacks. Moreover, environmental protection is a potential drawback hidden in the software aspect, which appears to be enough qualified. Synthetithecally, the defect in population urbanization, social urbanization and environmental protection leads Xiangyang to the situation of disequilibrium & half-integration, while Xiangyang could have achieve better one. Xiangyang is supposed to reversal the drawbacks in the quality of urbanization, through increasing efforts on environmental protection, creating more comfortable living environment, and improving life service. In addition, the feature of high level talent aggregation should be taken into account during all the developing guidance (Fig. 10.10).

Suizhou's dominance drawback is software aspect, of which there are two key elements, staff education structure, and culture and activities. And its potential drawback is technology & innovation, which is among integrally well-developed hardware aspect. Synthetithecally, the lack of innovative and advanced technology, the defect in education structure and cultural service, restrict the integration of Suizhou. The feature has correlation with the history of it. As the base of specialized automobile production, Suizhou high-tech zone has been economic development zone before the year of 2014, and it was finally be promoted as national level in 2015. The strength and output of specialized automobile production contributed to the hardware aspect in leading position. At the opposite, the long-term neglecting on software environment construction gave rise to the specific drawbacks as mentioned. Suizhou is supposed to pursue industrial structure upgrading and reforming, promote high-tech technology improvement, and develop research service. Only when the high-tech environment enhances, including both the enterprises' function and employees' structure, Suizhou could move towards industry-city integration (Fig. 10.10).

10.5 Concluding Remarks

We firstly make a comprehensive review of industry-city integration in China, and confirm the significance and value of our research. Then we refer to a scientifically tested evaluation system set up by Wang Xia, et al., and try to evaluate the samples in Hubei province, in the perspectives of the composite dimension, hardware foundation, software environment, and interactive dimension. According to the degree and status of industry-city integration, the samples are classified into 5 patterns: the pattern of equilibrium & integration (Wuhan), the pattern of equilibrium & half-integration (Xiaogan), the pattern of fluctuation & integration (Yichang), the pattern of fluctuation & not-integration (Jingmen and Xiantao), and the pattern of disequilibrium & half-integration (Xiangyang and Suizhou). Finally, we propose targeted strategies. We consider Yichang to follow the strategy of accelerating optimization to eliminate its potential drawbacks; while Xiangyang and Suizhou to follow the strategy of progressive adjustment to reverse their dominance drawbacks. To be more specific, Yichang should recover the contradiction between supply and demand, upgrade the satisfaction of interaction environment, through public participation and dynamic evaluation mechanism. Xiangyang should put efforts on environmental protection and improve life service, especially for high-level talent aggregation. Suizhou should promote industrial upgrading and reforming, and develop research service. Beyond that, Xiaogan, as the pattern of equilibrium & half-integration, is supposed to maintain the current development, will probably reach higher level of industry-city integration. Jingmen and Xiantao, as the pattern of fluctuation & not-integration, will need problem-oriented optimization or adjustment in the process of future development.

During the national-wide transformation era, the approach of analysis process from evaluation to strategies in Hubei case provides reference value to the midwest region of China. This research conducts a multi-sample horizontal analysis based on single-year statistics, limited to data acquisition. Considering industry-city integration is a long-term and comprehensive process, we hope to add process research and longitudinal analysis afterwards.

Acknowledgements We would like to give our special thanks for the support of the following institutions: National Natural Science Foundation of China (Program Number: 51478199).

References

Chang, Y. (2014). *Research on the development strategy of industrial park*. Shanghai People's Publishing House.

Chang, L., Xinyang, L., & Xiaoqiang, H. (2012). Path toward city-industry integration in new urban development zones. *Urban Planning Forum, 2012*(S1), 104–109.

Daogang, Z. (2011). The new idea of industry-city integration. *Decision-Making, 2011*(01), 1.

Dong, O., Heping, Li., Lin, L., Sidong, Z., & Yuan, Z. (2014). Industry-city integration in urban transition: Sino-Malaysia industrial park case. *Planners, 2014*(06), 25–31.

Du, B. (2014). Multiple analysis of industry-city integration. *Planners*, (06), 5–9.

Fu, J. (2015). *Shenyang economic zone industrial and urban integration development research*. Shenyang University of Technology.

Gangbiao, G. (2011). The spatial development of industrial clusters from the perspective of industry-city integration research. Zhengzhou University.

Gaobang, L. (2007). Research on the new index of measuring the relative level of the urbanization and industrialization. *The Journal of Quantitative & Technical Economics, 2007*(01), 46–55.

Hua, L. (2011). Discussion on the urban and industrial integration in the Suburb of Shanghai—taking Qingpu new city for example. *Shanghai Urban Planning Review, 2011*(05), 30–36.

Huadong, J. (2012). Discussion on the mutual compatibility of industry-city integration and urban construction: Taking Tianfu new district of Sichuan province as an example. *Reform of Economic System, 2012*(6), 43–47.

Jiang, Y. (2014). *Studies on modes and evaluation system of urbanism and industry integraion of provincial development zones*. Nanjing University.

Jing, L., Qian, G., & Yan, W. (2012). Industry-city integrate development oriented high-tech district development and planning strategies: Jinan East high-tech district case. *Planners, 2012*(04), 58–64.

Li, G. (2014). *The study of the development path of the integration of industries and cities in China*. Anhui University.

Li, Z. (2015). *Aresearch on the integration of industry and city in industrial cluster in Henan province: Based on the high-tech development zone in Zhengzhou*. Henan University.

Lin, S., Bin, G., & Xue, L. (2013). Fuzzy AHP evaluation on production-city integration of Shanghai zhanaiiang high-tech zone. *Industrial Technology & Economy, 2013*(07), 12–16.

Liu, C., Li, X., Hang, X. (2012). Path toward city-industry integration in new urban development zones. *Urban Planning Forum, 2012*(S1), 104–109.

Rongzeng, L., & Shuhua, W. (2013). Industry-city integration in urban new district. *Urban Problems, 2013*(06), 18–22.

Tang, X. (2014). *Research on the layout of the industrial parks in Shanghai and their integration with new town's development*. East China Normal University.

Wang, F. (2014). On the evaluation of industry and city fusion in industrial agglomeration area based on combined weight and four quadrant method. *Ecological Economy*, (03), 36–41+46.

Wang, X., Wang, Y. H., Su, L., Guo, B., & Wang, S. W. (2014). Index evaluation system on the degree of production-city integration in high-tech zones in China: Based on factor analysis and entropy-based weight. *Science of Science and Management of S. & T*, (07), 79–88.

Wenbin, L., & Hao, C. (2012). Analysis of city-industry integration and planning strategies. *Urban Planning Forum, 2012*(7), 99–103.

Yishao, S. (2016). A research on industry-city integration: Review and new exploration. *Urban Planning Forum, 2016*(05), 73–78.

Yongming, S., Xiaohua, C., & Jinlong, C. (2013). A review of the research on the industry-city integration in China based on spatial planning. *Journal of Chizhou University, 2013*(06), 77–80.

Zhu, Y. (2015). *Research of space optimization of deveopment zone based on industry and city integration—take Suzhou high-tech zone as an example.* Suzhou University of Science and Technology

Part II
Rural Planning and Urban-Rural Coordination

Chapter 11
Urban–Rural Coordination in Shaoxing: Small and Medium Town Development amid Urban–Rural Relationship Transformation

Wenting Jiang, Jian Liu, and Xiaoxuan Lin

11.1 Introduction

Urban–rural spatial relationship is one of the important issues in the discipline of urban and rural planning. According to the objective laws, urban–rural spatial relationship shall generally go through the development stages from differentiation to coordination along with the gradual improvement of industrialization and urbanization levels. Since the 1950s, China has for a long time seen an imbalanced urban -rural development and an obvious urban–rural differentiation under the joint effects of the socio-economic system, strategic policy orientation, and other factors, which have seriously restricted its sustainability of socio-economic development. In view of China's current urbanization process and socio-economic development, guiding the transformation of urban–rural relationship from differentiation to coordination is key to wholly improving the quality of urban–rural development and that of residents' life.

In the context mentioned above, a major strategy for coordinated urban–rural development was put forward, aiming to guide the formation of a balanced urban–rural development pattern, by giving full play to the supporting role of industry for agriculture and urban areas for rural areas, for the first time at the 16th CPC National Congress at the end of 2002. Since 2013, Xi Jinping, General Secretary of the CPC Central Committee, has repeatedly stressed the importance of improving the quality of urbanization and pushing forward the integration of urban–rural development at many important conferences, such as China's Central Government Urbanization Conference. He proposed to plan as a whole urban and rural areas, to promote the coordination in such aspects as production factor allocation, industry development, public services, ecological protection, etc.

W. Jiang · J. Liu (✉) · X. Lin
School of Architecture, Tsinghua University, Beijing, China
e-mail: liujian@tsinghua.edu.cn

© Springer Nature Switzerland AG 2021
W. Li et al. (eds.), *Human-Centered Urban Planning and Design in China: Volume I,*
GeoJournal Library 129, https://doi.org/10.1007/978-3-030-83856-0_11

The small and medium towns are key to the transformation of urban–rural relationship from differentiation to coordination, due to their unique location in urban-rural space and their unique position in urban system. In the transitional zones between severely differentiated urban and rural areas, small and medium cities and small towns emerged in large numbers and gradually became an important part of China's urban system, as China's planned economy was transformed into the market economy after its reform and opening up. At present, small and medium cities in China refer to the designated cities with an urban population of less than 1,000,000. Specifically speaking, small cities refer to those with a population of less than 500,000 in their urban areas, and medium-sized cities refer to those with a population of 500,000–1,000,000 in their urban areas.[1] Meanwhile, small towns refer to the sizable settlements chiefly engaged in non-agricultural activities which are different from both cities and villages, including designated towns and relatively developed rural market towns. Considering both the goal and path for urban–rural relationship transformation, the small and medium towns in this paper include the small and medium cities and small towns mentioned above.

Shaoxing, located in the Yangtze River Delta which boasts about the most developed economy and the highest urban agglomeration degree in China, has a strong economic base, stable social development, long development history, multiple local cultures, beautiful natural environment, and rich historic heritages (see Fig. 11.1). With an average town density of about 9.5 /1,000 km^2, it is a very typical small and medium town gathering area with comprehensive development advantages, i.e. the basic conditions for exploring the development mode of "urban–rural coordinated development and integral improvement" and the great potential to become an advanced demonstration area for urban–rural relationship transformation in China. By applying multiple research methods, including site survey, qualitative analysis and interdisciplinary research, this paper researches how to understand the transformation of urban–rural relationship and what role small and medium town development plays in the urban–rural relationship transformation, taking the urban–rural coordination in Shaoxing as an example.

11.2 Value Orientation in Urban–Rural Relationship Transformation

Since the reform and opening up, with the rapid economic development, China has witnessed an increasingly close urban–rural economic relationship, and most previous studies have described, evaluated, and classified urban–rural relationship from the perspective of economic development. Although the economic development indicator is very important to measure urban–rural development and urban–rural relationship, it should not be considered as the only standard. As China's economic

---

[1] *Notice of the State Council on Adjusting the Standards for Categorizing City Sizes (No. 1 [2014] of the State Council).*

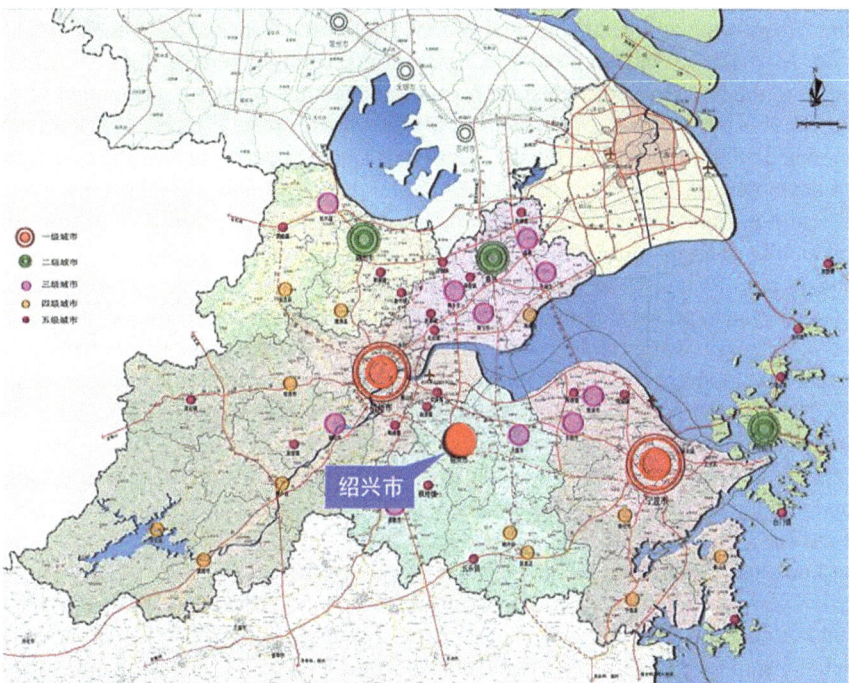

Fig. 11.1 Location of Shaoxing in Hangzhou Bay

development presents the new normal of "quality and efficiency", it will no longer have a dramatic incremental urban development. Accordingly, both urban–rural space development demands and patterns will change, and urban–rural relationship should not be understood only from the perspective of economic development but to reflect the change from economy-oriented to quality-oriented. When the urban–rural relationship undergoes the transformation from binary opposition to mutual coordination, it is necessary to pay more attention to the comprehensive quality improvement of economy, society, culture, environment, and so on.

11.2.1 Definition of Urban–Rural Relationship

The transformation of urban–rural relationship follows a certain objective development law. According to the experience of Western developed countries, it can be divided into the following five development stages1: ① Urban–rural symbiosis period, from the birth of villages and cities to the 10th–thirteenth century. In this period, with the development of human settlements, villages came into being. With the social division of labor, a few villages gradually associated to create cities. At that time, there were almost no visible distinction between urban and rural areas in function, spatial form, industry, etc. ② Urban–rural separation period, from the Middle Ages to the beginning of industrial revolution. With the further development of the social division of laber, more and larger cities arose from villages. Cities accommodated more industry and business activities and took rural areas as the hinterland of development. Urban and rural areas had a relatively stable complementary relationship in function, environment, etc. ③ Urban–rural opposition period, from the beginning of industrial revolution to the take-off of post-industrial economy. The functional distinction between urban and rural areas became clearer and clearer. Cities began to play a leading role in regional development with economy and population increase, while rural areas were faced with underdevelopment. ④ Urban–rural equal development period, when a country sees a rapid economic development and its urbanization process tends to be stable. The population move back towards and increase in rural areas. There still exists a diminishing functional distinction between urban and rural areas, but a more obvious complementary relationship in ecology and culture. ⑤ Urban–rural integration development period, when a country sees a transformation from the advanced stage of industrial economy to the development of knowledge economy and a coordinated and stable national economy. There still exists obvious difference between urban and rural areas in function and spatial form, but both urban and rural residents have almost same living quality. The urban–rural relationship tend to be balanced and stable again.

Among the above five stages, equal development which is between space opposition and space integration, is an important stage of urban–rural relationship transformation. At this stage, the binary opposites between urban and rural areas are fully highlighted and need to be solved. Meanwhile, a relatively mature foundation of both the socio-economic development and urbanization level is conducive to urban–rural development.

11.2.2 Evolution of China's Urban–Rural Relationship After 1949

China was a large traditional agriculture-based country. In the early years after the founding of the P.R.C., in order to improve national comprehensive strength as soon as possible, China had taken economic development as an important goal, put more funds into industrial development, and thus to some extent neglected rural development and construction. At the same time, China's unique household registration system and the relatively backward urban–rural dual economic structure led to a serious urban–rural differentiation, causing a lot of obstacles on the road of urban–rural relationship moving on to equal development and space integration.

After the reform and opening up, China's urban–rural relationship slowly changed with the gradual relaxation of the planned economy system. However, at the same time, the problem of low income of farmers due to the low prices of food and other agricultural and sideline products, the lack of other income sources, and so on was gradually highlighted. With the social and economic development, the gap between urban and rural areas in such aspects as economic development, public service, facilities construction, and social security were gradually widened. Urban–rural dual economic structure and the Three Rural Issues (issues of agriculture, farmers, and rural areas) gradually became the main obstacles to China's socio-economic development and urban–rural relationship transformation.

According to the latest statistics, up to the end of 2016, China's urbanization rate was 57.35%, the proportion of the three sectors was 8.6: 39.8: 51.6, and annual GDP per capita was 53,980 yuan (about 7,841 US dollars).[2] Due to its vast territory, China's demographic and social development and economic and industrial development are imbalanced in space. From the macroeconomic statistics, it can be noticed that China's relatively developed areas have entered a transitional stage from urban–rural space opposition to equal development, having the practical demands for urban–rural relationship transformation and the basic conditions for urban–rural coordination. For instance, by the end of 2016, the average urbanization level of Shaoxing was 58.5%, the proportion of its three sectors was 4.5: 49.2: 46.3, and GDP per capita was 14,245 US dollars.[3] Judged from the comprehensive indicators, Shaoxing is at the transitional stage from urban–rural space opposition to equal development.

11.2.3 Value Orientation and Analytical Framework of Urban–Rural Coordination

The content of urban–rural coordination involves all aspects of social and economic development. In recent years, many scholars have studied the concept and connotation of urban–rural coordination, and put forward different understandings from the perspectives of different disciplines. From a sociological point of view, Yishao Shi (2003) believes that urban–rural coordination is to break the barriers between urban and rural areas, make urban–rural economic and social life link closely, and gradually narrow till eliminate the basic differences between urban and rural areas, ultimately to reach urban–rural integration. From an economic point of view, Xiaoming Wang et al. (2010) argues that urban–rural coordination is the objective requirement of the increasingly strong agriculture-industry tie in modern economy. They suggest enhancing the economic exchange and cooperation between urban and rural areas and rationally distributing productive forces, in order to achieve the best possible economic benefits. From the perspective of ecology and environment, Huimin Jia (2010) points out that urban–rural coordination should be integrated with urban–rural

---

[2] *National Economic and Social Development Statistical Communique of the People's Republic of China in 2016.*

[3] *National Economic and Social Development Statistical Communique of Shaoxing City in 2016.*

Table 11.1 The indicator framework of urban–rural relationship transformation

Relevant factors	Key indicators
Economy and industry	Economic aggregate, economic structure, enterprise composition; industry composition, leading industry, featured industry
Population and society	Urbanization level, employment level, income level, education level, social security level
Environment and ecology	Air quality, water quality, waste disposal, biodiversity
Space layout	Landscape pattern, land use, distribution of villages and towns, village-town system
Facilities	Municipal facilities (road, water, electricity, gas), service facilities (education, health care, culture, sports)
Cultural features	Historical heritage protection, local characteristics protection
Architectural landscape	Architectural style, village landscape, townscape, natural landscape
Institutional construction	Related policy measures and their implementation mechanism

ecological environment, to promote healthy and coordinated urban–rural development. From the perspective of space planning, Yishao Shi (2013) thinks that urban–rural coordination refers to taking cities and countryside as an organic whole and optimizing the spatial layout and strengthen the functional complementation.

Based on the above mentioned research viewpoints, in terms of a long-term goal, urban–rural coordination should be guided by the comprehensive development of quality rather than a single economic indicator. Therefore, efforts should be made to pursue the overall quality of urban–rural development, especially the improvement of residents' life quality and comprehensive happiness index on the basis of a moderate economic growth, in order to improve the attractiveness and competitiveness of urban and rural areas as a whole. In the process of urban–rural relationship transformation, people concerned should pay attention to the differences of urban–rural development in all aspects listed in Table 11.1, and make macro strategies and specific strategies for urban–rural overall planning and construction in future based on the corresponding indicator framework.

11.3 The Role of Small and Medium Towns in Urban–Rural Relationship Transformation

Development of small and medium towns is of great significance to urban–rural relationship transformation. From the perspective of quantity and scale, small and medium towns account for a large proportion in China's urban system, which is conducive to'promote the development of villages. From the perspective of spatial distribution, small and medium towns that are mostly located in the junction areas of large cities (metropolitan area) and villages are conducive for high-quality production factors to move to and from between urban and rural areas. Therefore, in order to promote the comprehensive development of urban and rural areas and the transformation of urban–rural relationship, development of small and medium towns and implementation of urban–rural coordination strategies need more attention.

11.3.1 Numerous Small and Medium Towns Are an Important Part of China's Urban System

According to the National New Urbanization Plan (2014–2020), China saw a dramatic increase in the number and size of cities (towns) at all levels from 1978 to 2010 (see Table 11.2), especially the number of designated towns. By 2010, there were 19,928 designated towns and rural market towns in total. Due to the quantity, small and medium towns occupy an important position in China's urban system, have an advantage to support the development of villages, and play an important role in urban–rural coordination.

For example, as a typical small and medium town gathering area, Shaoxing has a relatively high town density with an urban system consisting of a central city, four sub-cities (counties), 20 central townships, and 59 general townships. In particular, southern Shaoxing area comprises a large number of towns and higher town density (see Table 11.3), accommodating about 50% of the total population of Shaoxing, which is of great significance to facilitate urban–rural coordination.

Table 11.2 Changes in the number and size of cities (towns) in China from 1978 to 2010

Cities		1978	2010	Variation
	Cities with more than 10 million population	193	658	365
	Cities with 5 to 10 million population	0	6	6
	Cities with 3 to 5 million population	2	10	8
	Cities with 1 to 3 million population	2	21	19
	Cities with 0.5 to 1 million population	25	103	78
	Cities with less than 0.5 million population	35	138	103
Designated towns		2,173	19,410	17,237

Table 11.3 Number of towns and town density in southern Shaoxing area

	Number of towns	Town density (per 1,000 km^2)
Yuecheng District	7	19.8
Zhuji City	23	10.0
Shengzhou City	11	8.4
Xinchang County	8	6.1

11.3.2 The Geographical Locations of Small and Medium Towns Make Them a Spatial Link Between Urban and Rural Areas

In terms of spatial distributions, small and medium towns are usually located at the inner-edge or periphery area of metropolitan regions, close to rural areas. Driven by the development of large cities, small and medium towns have a certain basis for industrial development and infrastructure construction, yet with the advantages of lower construction density, better ecological space, and more natural scenery. For the central city, small and medium towns can accept its population and function (outward) relocation and provide a higher quality of life with fresh air, clean drinking water, quiet living environment, and lower living cost. For surrounding rural areas, small and medium towns can provide job opportunities and public services, etc. Therefore, the numerous and widely distributed small and medium towns play an important role in linking urban and rural areas, as well as promoting urban–rural respective advantages complementary to each other and the production factor flows between urban and rural areas.

At present, there has formed a metropolitan development pattern in the plain area of northern Shaoxing, whereas towns in southern Shaoxing are relatively scattered due to numerous hills and mountains (see Fig. 11.2). At the same time, Shaoxing also has over 20 provincial-level key townships covered by the 12th Five-Year Planning Outline of Zhejiang Province, including Gaobu, Qianqing, Yangxunqiao, Pingshui, Datang, Diankou, Songxia, Zhangzhen, Fenghui, Changle, Ganlin, Ruao, Lanting, Fuquan, Fengqiao, Paitou, Ciwu, Xiaoyue, Huangze, and Chengtan (see Fig. 11.3).

Fig. 11.2 Spatial distribution of towns in Shaoxing

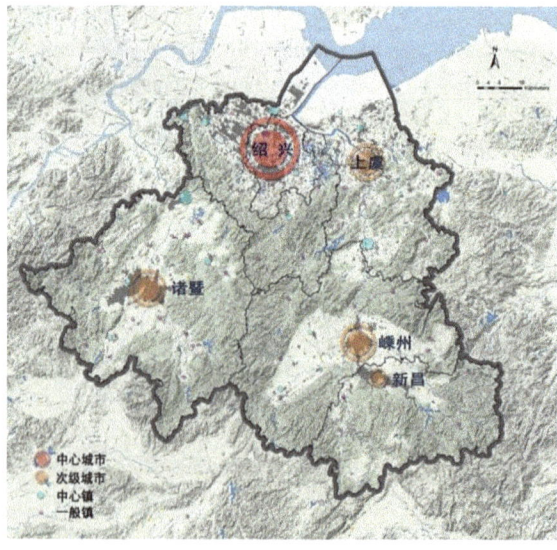

Fig. 11.3 Distribution of
provincial-level key
townships in Shaoxing

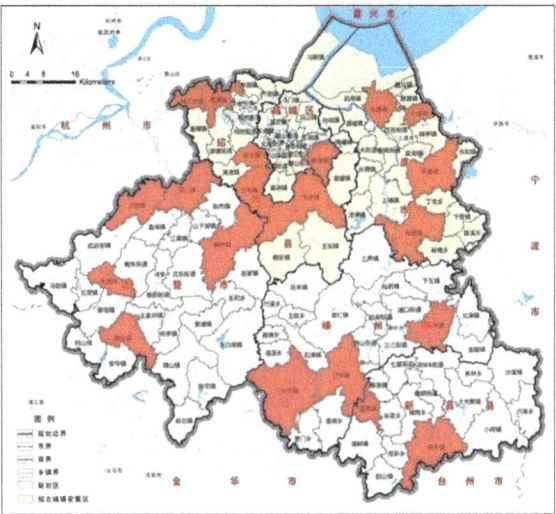

Scattered small and medium towns, in particular the key townships, are of great significance in promoting the urban–rural coordination in Shaoxing.

11.3.3 Small and Medium Towns Are Important Carriers for the Implementation of Urban–Rural Coordination Strategies

The goal of urban–rural relationship transformation is to achieve the coordinated development of urban and rural areas. Urban–rural coordination is not simply to change villages into cities, but to promote the overall development level on the basis of preserving the characteristics of cities and villages and to let both rural and urban residents have access to equal rights, equalized public services, and homogenized life quality. The ideal state of urban–rural coordination is: towns at all levels can develop into regional or local centers with healthy economy, sound urban functions, and balanced social securities, as well as the gathering places for the sustainable development of featured industries and local rural population transfer, basically reaching the urban development level of developed countries; rural areas can develop into the model of new socialist countryside with beautiful scenery, prosperous economy, comfortable living environment, and distinctive features, basically reaching the rural development level of developed countries.

The key to achieve urban–rural coordination is the supporting role of urban areas for rural areas. As an important part of urban system and an important spatial link between urban and rural areas in China, small and medium towns cannot be ignored in urban–rural coordination planning strategies. In the long-term development process, most of the small and medium towns will have a certain development basis in the aspects of economy, society, facilities, etc., but there also exist some problems, such as imbalanced economic growth and ecological & resource protection, lack of historical and cultural protection, unreasonable spatial structure, and lower quality of development. Therefore, relevant planning strategies should be proposed and implemented at the township level, in order to play the driving role of small and medium towns to the surrounding villages and to achieve urban–rural coordination.

For example, the urban areas of Shaoxing are able to play a supporting role for rural development as they have strong economic strength, developed private economy, distinctive industry characteristics, relatively intensive population, higher level and stable process of urbanization, sound social security, relatively complete public service and infrastructure construction. Thanks to the outstanding characteristics, the superior resource endowment, and unique development advantages in the environment & ecology, industry features, local culture, and so on of rural areas, Shaoxing has the basis and potential to comprehensively improve the development quality of rural areas. At the same time, there also exist problems in Shaoxing, such as the big gap between urban and rural areas in terms of public service facilities construction, a certain degree of pollution in part of the water areas, the shortage of historical and cultural elements excavation, relatively poor rural architectural landscape. The key to realize urban–rural coordination in Shaoxing is to formulate urban–rural coordination planning strategies at the township level, so as to fully play and strengthen the respective resources and basic advantages of urban and rural areas and reinforce the complementary support of urban and rural areas.

11.4 Strategies and Suggestions for Urban–Rural Coordination

Urban–rural coordination aims directly at a moderate economic growth rate and all-around improvement of the development quality of urban and rural areas, and it may further regulate the distribution of population, space, resources, etc., so as to moderate the various problems caused by the socio-economic development in the process of urban–rural urbanization. From the viewpoint of population migration, there is a basic "marching forward" social development law in urbanization, making population migration be characterized by level-by-level or skip-level leaping forward from rural areas to small and medium towns, counties, small and medium cities, even large cities and mega-cities, which brings about such issues as rural hollowing and overcrowded large cities. Improving urban–rural life quality by way of urban–rural coordination can offer people more diversified migrating choices, which is of great significance to wholly improve China's socio-economic development level and urbanization quality. Therefore, the planning strategy for urban–rural coordination should focus on improving the comprehensive development level and attractiveness of small and medium towns. The township-level administrative units, accounting for the largest proportion, should be taken as important implementation units for urban–rural overall planning strategies.

11.4.1   Uncovering and Fostering Featured Township Industries

Fostering the featured industries of small and medium towns is a cornerstone for urban–rural coordination. First of all, the industry development shall immediately bring about the improvement of the economic level, providing a strong economic foundation for small and medium towns to promote the development of social causes, to improve infrastructure construction, to strengthen ecological environment management, and to ultimately improve comprehensive development level. Secondly, directly facing the vast rural areas, small and medium towns that have a good industrial development can have a strong employment capacity for rural surplus laborers. Reciprocally, rural areas can also be the strong bearer of industry transfer on the basis of its considerable human and land resources. This kind of urban–rural interaction can help to form an interworking industry system between small and medium towns and rural areas, providing stable job positions and attracting more local talents to stay at their hometown. Thirdly, enhancing the development of featured industries may help small and medium towns to form a distinctive characteristic, so as to improve their comprehensive competitiveness and attractiveness. Therefore, featured industries are the foundation of small and medium town development, and uncovering and fostering featured industries are the first and foremost step for the urban–rural coordination of small and medium towns.

Shaoxing area, for example, has a traditional advantage in economy and industry development featured by "each town and village with an industry". Among others, Shengzhou City began to develop the tie industry from the mid-1980s, and now has become the "centers" of production and processing, wholesale sales, product display, information services, research and development, quality testing in the tie industry; Shanxiahu Township, known as a "Chinese Pearl Town", is the largest breeding place and processing and trading base of freshwater pearl in China. Distinctive industries have contributed to the strong economic strength of Shaoxing and laid a solid economic foundation for urban–rural coordination. In recent years, Shaoxing has accumulated more advanced experience in the aspects of exploring featured small township construction by making full use of the advantages of local industry characteristics. For instance, the socks township located in Zhuji City is a featured town based on the socks industry of Datang Township, which has expanded out to some functional areas with socks industry as the theme, such as smart manufacturing, cultural tourism, entrepreneurship innovation etc. This has enriched the business formats, driven the development of relevant industries in rural areas, and improved the comprehensive development level and urban competitiveness.

11.4.2 Strengthening and Equalizing Public Services and Infrastructure

It is an inevitable requirement for urban–rural coordination to promote the equity of public service facilities and infrastructure in urban and rural areas. The equity level of public service and social security in urban–rural areas can be gradually improved through constantly encouraging public service facility networks of small and medium towns to extend to and cover rural areas and promoting the construction of public service facilities in rural areas, so as to reduce urban–rural residents' differences in life quality. The spatial link between urban and rural areas can be strengthened through advancing the construction of regional transport networks and developing public transport, to promote the production factor flow between urban and rural areas. The entire supporting capacity of the infrastructure system can be improved through enhancing the infrastructure system construction, to provide the hardware support for urban–rural coordination. Consequently, it is an essential requirement for urban–rural coordination to comprehensively boost the urban–rural coordination and equity of public services and municipal infrastructure.

Taking Shaoxing as an example, the public service facilities and municipal transport infrastructure construction in Shaoxing are relatively complete, but there are also issues, such as relatively weak grassroots cultural facilities construction, relatively large gap between urban and rural public service facilities, relatively low degree of smooth contact between rural areas and outside places, and inadequate waste and sewage treatment capacity. Targeting at the above issues, Shaoxing shall take a long time to optimize and improve urban–rural basic public services and infrastructure, which will be regarded as the focus of improving people's livelihood. It is suggested to take such initiatives as increasing the investment in the facilities and improving the social security mechanism to narrow the gap between urban and rural areas in terms of comprehensive service levels, so that more urban–rural residents can have access to better education, more convenient medical care, richer entertainments, and more stable life.

11.4.3 Formulating Township-Level Development Strategies

A comprehensive and systematic multi-disciplinary development strategy is an important guarantee for urban–rural coordination, which aims to achieve a relatively balanced urban–rural development in social, economic, environmental, population, and other aspects. Therefore, at the township level, in addition to developing featured industries to consolidate the economic base and improving the comprehensive service capacity through promoting the equalization of supporting facilities, it should include the planning strategy for population and society, environment and ecology, spatial layout, cultural characteristics, architectural landscape, institutional improvement, and so on, so as to systematically develop a strategy system for urban–rural coordination.

For instance, located at the foot of Dongbai Mountain, Dongbaihu Township in Shaoxing boasts about its excellent natural resources and ecological conditions, featured folk architecture, relatively complete yet large urban–rural disparity in terms of public services and infrastructure, and outstanding agricultural and industrial base. Based on the analysis of current conditions and resource characteristics, urban–rural coordination strategy of Dongbaihu Township should focus on the comprehensive promotion of economic industry, cultural characteristics, spatial layout, the environment and ecology, institutional improvement, and so on (see Table 11.4).

Table 11.4 Shaoxing Dongbaihu township urban–rural coordination planning strategy system

Related aspects	Strategies and suggestions
Economic industry	On the basis of strengthening the existing advantages of agriculture and industry, the tourism services industry with agritainment as the representative shall be developed relying on major tourist areas
Cultural features	Strengthen the utilization of existing cultural advantages, such as ancient folk building complex in Sizhai Township, and create a unique Dongbaihua Township cultural brand
Spatial layout	Avoid large-scale migration and concentration of villages, maintain as much as possible the existing village texture and the relationship between villages and farmland & landscapes on the basis of a comprehensive coverage of the infrastructure, making the village a part of the landscape
The environment and ecology	Strengthen the protection of natural resources and the supervision of existing industrial enterprises, put an end to industrial pollution, guide the transformation of agriculture, encourage the development of ecological agriculture, reduce agricultural pollution, enhance the village hygienic condition remediation, encourage the use of renewable energy, promote beautiful countryside construction
Institutional improvement	Provide financial support for the development of the tourist area, provide appropriate subsidies for the renovation of villagers' housing, establish a new housing construction standard system, provide a series of technical support for villagers from such aspects as architectural modeling, building materials selection, construction methods

11.5 Conclusion

It takes a long period to achieve transformation or urban–rural relationship. During this process, a right value orientation contributes to make the transformation scientific, efficient and orderly. As China's economic development presents the new normal, an overall quality-oriented value is relatively scientific for urban–rural relationship transformation.

It is one of the most important approach to urban–rural coordination that exerting the leading role or urban area to rural area. In China, there are large numbers of small and medium towns, which account for a large proportion in China's urban system and play a role of space link between urban and rural areas. Therefore, development of small and medium towns is of great significance to urban–rural relationship transformation. Taking the planning practices in Shaoxing district as an example, this paper may provide reference for urban–rural coordination planning in similar districts where small and medium towns gather.

Based on studies about Shaoxing case, this paper comes up with three strategy suggestions. The first one is to uncover and foster featured township industries as the foundation of small and medium towns. The second one is to strengthen the equity of public services and infrastructure. The last one is to formulate system framework of township-level development strategies.

In conclusion, the key to urban–rural coordination is to promote the life quality of rural residents and to narrow the gap between urban and rural areas. However, it is worth noting that the ideal goal of urban–rural coordination is not simply to change villages into cities, but to make rural residents have the same rights, public services, life quality and happiness feelings as urban residents.

Acknowledgements This study is supported by the research project of Land Property Rights, Land Readjustment & Rural Planning: A Study on Implementation Mechanism of Integrated Utilization of Rural Collective-Owned Constructible Land (No. 51678326), sponsored by National Natural Science Foundation of China.

References

Chenery, H., Robinson, S., Syrquin M. (1986). Industrialization and growth: A comparative study. Oxford University Press.

Cheng, M. (2007). *International experiences from the development of small and medium cities and views on development of small and medium cities in China [D]*. Wuhan: Central China Normal University.

Jia, H. (2010). *Research on ecology landscape in the process of urban-rural coordination [D]*. Baoding: Hebei Agricultural University.

Li, W. (2015). The developmental experience for reference of development of small and medium cities and small towns. *Journal of Shanxi Agricultural University (Social Science Edition), 14*, 175–179.

Lijuan, Wu., & Liu, Y. (2012). Hui Cheng. Review on the dynamic mechanism and key elements of urban-rural integrated development in China. *Economic Geography, 32*, 113–117.

Liu, J. (2012). *Research on urbanization development mode of china's medium-sized cities and small cities [D]*. Chongqing: Chongqing University.

Shi, Y. (2003). Theory and practice of urban-rural integration: Review and analysis. *Urban Planning Proceedings, 2003*(1), 49–54.

Shi, Y. (2013). New-type urbanization and small town development in China. *Economic Geography, 33*, 47–52.

Wang, Z. (2000). *Urban-rural spatial integration theory: A system research on urban-rural spatial relation during the process of China's urbanization sustainable development*. Shanghai: Fudan University Press.

Wang, X., & Zhang, M. (2010). The key to urban-rural coordination is the development of industry in towns: Analysis based on Ande Town Chengdu City. *Rural Economy, 4*, 37–41.

Chapter 12
Spatial Growth of Urban and Rural Construction Land and Policy Impact Mechanism in Hangzhou

Haiyan Pang and Yonghua Li

12.1 Introduction

With the rapid advance of urbanization, spatial growth of urban and rural construction land and urban spatial morphology evolution become the inevitable trend of urban development (Yao et al., 2014).Yet urban space transition also brings the spread of urban expansion and other issues (Mingyue, 2012). Big cities in China will gradually change from incremental expansion to less-growth and renewal in the future (Wenyu, 2014; Yonghong, 2005). Cities are complicated synthesis that their space evolution is under the dual role of the market and government policy (Jingxiang et al., 2013; Song, 2007; Wang et al., 2012). Related to the influence of spatial policies, this paper summarizes the characteristics of urban spatial growth and in the temporal scale to reveal the significance of the spatial policy mechanism.

In the recent studies of other scholars, the main public policy instruments are identified and briefly described, including policy implications and classification (Doern and Pal 1988), policy implementation systems (Cheng, 2015; Dennis Wei & Liefner, 2011; Dikeç, 2002), and policy performance evaluation (Salet & Thornley, 2007). Connotation and classification of public policy are classified and summarized according to the classification criteria such as policy-making part, policy function, and policy proceeding period (Jian & Yu, 2008).

In the policy implementation system, existing research focuses on the connection between public policy and planning management, and has gradually established a top-down implementation system and a bottom-up feedback mechanism (Cheng, 2008; Ming & Chen, 2013; Shiying et al. 2017). On the policy performance evaluation,

H. Pang · Y. Li (✉)
Institute of Urban and Rural Planning Theories and Technologies, Zhejiang University, Hangzhou 310058, China
e-mail: lyh_zju@126.com

H. Pang
e-mail: 876798251@qq.com

© Springer Nature Switzerland AG 2021
W. Li et al. (eds.), *Human-Centered Urban Planning and Design in China: Volume I*,
GeoJournal Library 129, https://doi.org/10.1007/978-3-030-83856-0_12

space policy measure, index, and evaluation system are already established (Chang et al., 2017; Chenfei & Qun, 2015; Jiansheng et al., 2014; Li & Yilun, 2017), such as setting land policy scenarios in future urban space (Jiansheng et al., 2014).

For the research of urban spatial form evolution, the existing literature mainly uses the extension strength retention rate statistics, the transfer matrix of land use and landscape components (Lihua et al., 2014; Wei et al., 2017; Wenze et al., 2013) as analysis tools, and uses quadrant analysis and spheres as analysis methods (Congjian et al., 2013; Liang, 2016) to form the urban spatial structure and inductive mode (Lihua et al., 2014; Wenze et al., 2013). Recent public policy studies focus on policy content, and the quantitative research to measure policy effects, but lack discussion of the impact of policy as a mechanism. There are few researches of urban spatial evolution analyzing in a systematic policy perspective. The related studies above seldom think about the internal logic of growth policy and space correlation analysis.

Taking Hangzhou as an example, this paper uses the sector analysis, the expansion intensity index, and the centroids of urban and rural construction land to analyze the spatial growth from 2000–2015. The results are then related to spatial policy to reach the rules of space evolution form and reveal the policy mechanism.

12.2 Research Area and Method

12.2.1 Research Area

Hangzhou is located in the delta of Yangtze River in China, south of the Beijing-Hangzhou Grand Canal. Hangzhou is the capital of the Zhejiang province and has a high status in economic, culture, science and education in the Zhejiang province. Since the twentieth century, Hangzhou has experienced four administrative changes: Xiasha adjusted from the Yuhang District into the Jianggan District in 1993, then three towns from the Xiaoshan District departed out to the establish Binjiang District in 1996; Xiaoshan and Yuhang adjusted to new districts of Hangzhou city in 2001; Fuyang adjusted to a new district of Hangzhou city in 2015; Linan adjusted to a new district of Hangzhou city in 2017. Along with these administrative division adjustments, Hangzhou surpassed Shanghai and Nanjing to become the largest city in eastern China. The research scope of this study is mainly 3068 Km² of eight districts including Shangcheng, Xiacheng, Gongshu, Jianggan, Xihu, Binjiang, Xiaoshan and Yuhang District. The time span is from 2000 to 2015, and the research nodes are intercepted every five years.

12.2.2   Data Sources

Data source includes the scope of Hangzhou Landsat TM/ETM + image data
(October 11, 2000, November 26, 2005, September 21, 2010, and August 2, 2015;
track number 119,39). Data preprocessing steps include atmospheric correction,
geometric correction, and administrative boundary image cropping. Based on the
classification standard of construction land, woodland and garden plot, cultivated
land, wetland, water standard, and by using a support vector machine in RS platform
(Fig. 12.1), the four kappa coefficient are all better than 90% (2000, 0.942; 2005,
0.953; 2010,0.947; 2015, 0.942). Data source also includes space policy since 2000 in
Hangzhou (Table 12.1) such as government gazette files and department documents
(http://www.hangzhou.gov.cn/).

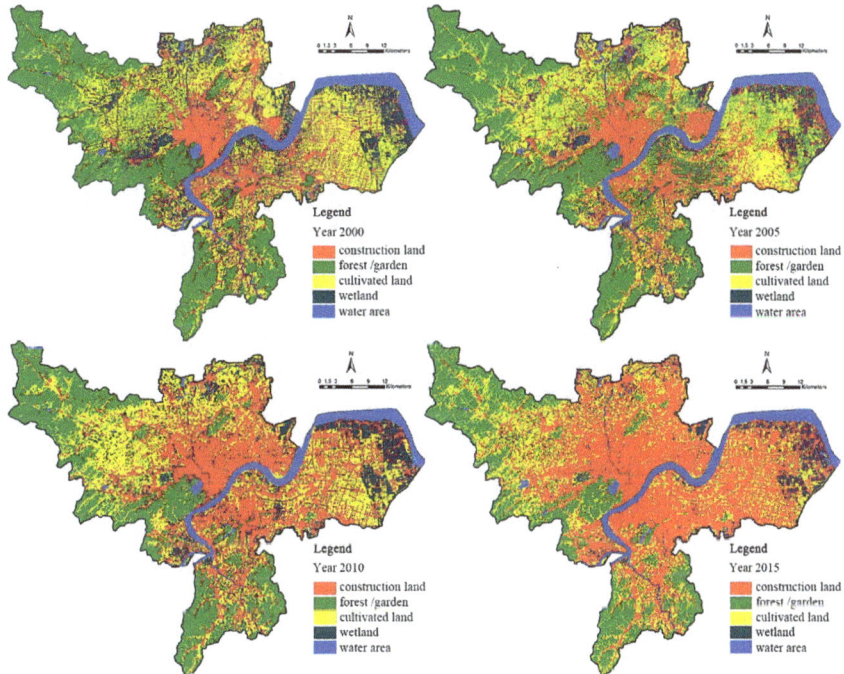

Fig. 12.1   Hangzhou supervision classification from 2000 to 2015

Table 12.1 Hangzhou space police summary in 2000–2015

Policy type	2000–2005	2005–2010	2010–2015
Strategic policy	• The "West Lake Ara" towards the "Qiantang River Era" • Set up Xiaoshan District, Yuhang District	• Promote the financial center in the south wing of Yangtze delta • Accelerate the construction of new center of Qianjiang	• Set up Fuyang District • Speed up the planning and construction of characteristic towns
Industrial policy	• Set up the economic and technological industry platform of Xiasha • Set up a national high-tech industry platform of Binjiang	• Set up the Qianjiang science and technology zone • Set up Xiaoshan economic and technological development zone	• Establish the east industrial agglomeration area • Set up a great corridor of the west section in Hangzhou • Ten major industries development plan in Hangzhou
Planning policy	• Development zoning plan in Binjiang District	• Hangzhou city master plan (2001–2020) • Concept plan of Qianjiang science and technology zone	• Concept plan of future science and technology city
Transportati policy	• Accelerate the construction of " westward traffic "	• Speeding up subway construction	• Speeding up the development of urban rail transit construction
Ecological policy	• Western region protection and development planning • Overall planning of Xixi national wetland park	• Concept plan of ecological belt in hangzhou • Hangzhou green space system planning	• Promote the protection and construction of the ecological landscape on both sides of the Yangtze river

12.2.3   Methods

The Sector Analysis

In the ArcGIS10.2 platform, this paper takes Wulin Square as the center with a radius of 50 km from the north by east 12.25° to divide the research area into 16 equal angles. Then, 16 different segments of the construction land are extracted. Using a radar map, the increment and expansion intensity of each sector can be shown in different time periods.

The Construction Land Expansion Intensity Analysis

The extended strength index is suitable for the analysis of the expansion state, which can accurately reflect the speed and intensity of the expansion of urban and rural construction land in different periods. The expression of the extended strength index is:

$$E_i = \frac{U_b - U_a}{U_a} \times \frac{1}{T} \tag{12.1}$$

In the above formula, E_i refers to the urban expansion intensity index, and U_a and U_b respectively represent the total area of construction land during the period of a and b, and T represents the time span from a to b, generally in years.

The variation index is used to analyze the differentiation degree of the expansion intensity, which can directly reflect the differentiation of the spatial and temporal evolution of construction land in different periods. The expression of the coefficient of variation is:

$$C_v = \frac{\sigma}{\mu} \tag{12.2}$$

$$\sigma = \sqrt{\frac{1}{N} \sum_{i=1}^{N} (E_i - \mu)^2} \tag{12.3}$$

$$\mu = \frac{1}{N} \sum_{i=1}^{N} F_i \tag{12.4}$$

In the above equation, C_v represents the variation index, and E_i is the expansion intensity index of the construction land of azimuth i, and N is the number of sector fields.

The Construction Land Centroids Transfer Analysis

Through the calculation of the distance and angle of the center of gravity of the construction land, it can reveal the structural change trend at different times. Calculation formula f construction land center:

$$x = \sum_{i=1}^{n} M_i x_i / \sum_{i=1}^{n} M_i \tag{12.5}$$

$$\bar{y} = \sum_{i=1}^{n} M_i y_i / \sum_{i=1}^{n} M_i \tag{12.6}$$

In the above equation, M_i represents the area of the urban construction land unit, and x_i and y_i represent the central coordinates of the i th unit respectively.

12.3 Result

12.3.1 Incremental Features of Urban and Rural Construction Land

The General Incremental Features

From 2000 to 2015, the construction land of urban and rural areas in Hangzhou increased relatively uniformly, the eastern part increased more while the southwest decreased. The urban and rural construction land area increased from 2526.04 to 3425.75 km^2, with a net increase of 899.71 km^2. Among them, the increment was the largest in e-azimuth, and the increment of SEE and NEE was followed by the least increment in the urban and rural construction land (Fig. 12.2-A).

The Period Incremental Features

From 2000 to 2005, the growth area of urban and rural construction land was 128.41 km^2, with a low growth rate of 5.08%. The urban and rural construction land grew in the east and the southeast, with the largest increment in E and the second in the SSE (Fig. 12.2-A1). From 2005 to 2010, urban and rural construction land increased 304.92 km^2, with a moderate growth rate of 11.47%. The urban and rural construction land increased on both the east and west sides, with a significant increment of NEE, E and SEE, and the least increase in SW and N (Fig. 12.2-A2).From 2010 and 2015, the growth area of urban and rural construction land was 466.38 km^2, with a high growth rate of 15.20%. Urban and rural construction land grew radially in the northeast, northwest, southeast and southwest, with a significant increment in E and SEE. In addition, the growth of SW azimuth land was still under control and rarely increases (Fig. 12.2-A3).

12.3.2 Expansion Features of Urban and Rural Construction Land

The General Expansion Features

For 2000 to 2015, the average expansion intensity of urban and rural construction land in Hangzhou was 0.33, and the variation coefficient was 0.31. Except for the

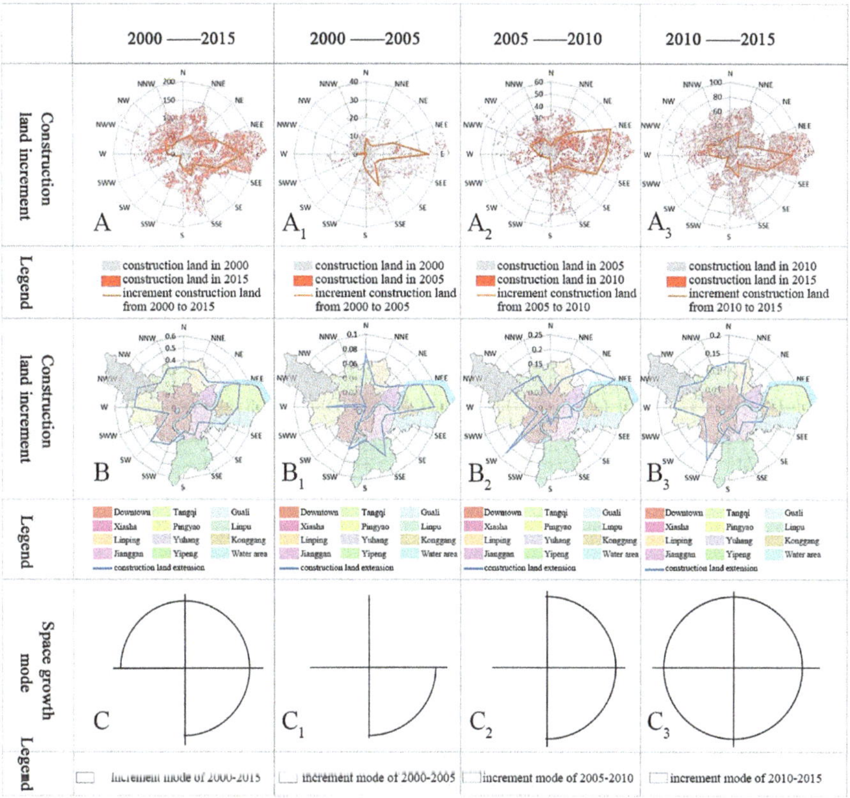

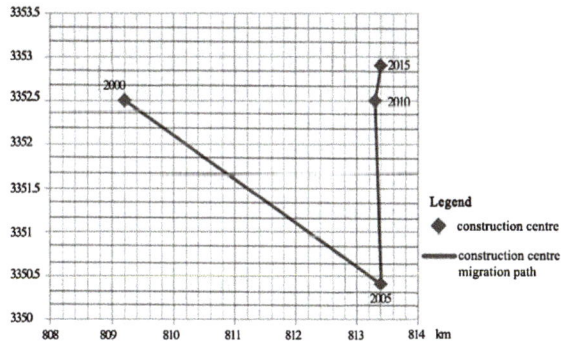

Fig. 12.2 Hangzhou incremental features and extension features in 2000–2015

Fig. 12.3 Hangzhou centroid shift characteristics in 2000–2015

stagnation of southwest expansion, the rest was balanced expansion. In particular, the NEE direction of urban and rural construction land expansion was the largest, followed by E and W expansion, with the smallest expansion intensity SWW and SE (Fig. 12.2-B).

The Period Expansion Features

From 2000 to 2005, the average extension strength of urban and rural construction land in each direction was 0.05, with a variation coefficient of 0.76. The expansion intensity was small though the extension strength gap was big, which present a disequilibrium mode of spatial growth in east and southeast (Fig. 12.2-B1). The direction of extension strength in 2005–2010 averaged 0.09, with a variation coefficient of 0.62. The extension strength disparity fell, which presented a partial equilibrium mode of agglomeration to the northeast and northwest (Fig. 12.2-B2). The direction of extension strength in 2010–2015 averaged 0.09, with a variation coefficient of 0.36, the extension strength difference fell further, which presented an equilibrium mode of spatial growth in northeast, northwest, southeast and southwest (Fig. 12.2-B3).

12.3.3   Center of Gravity Transfer Features

From 2000 to 2005, urban and rural construction land shifted a distance of 4.6 km and transfered an angle of 26.86 degrees. The emphasis had shifted from Kaixuan Street (west of Qiantang River) to century city (east of Qianjiang River). During this period, the center of city development shifted to the southeast. From 2005 to 2010, the center of gravity of urban and rural construction land shifted 2.4 km northward maintaining consistent growth in this direction, and the angle was 88.72 degree. From 2010 to 2015, the center of gravity of urban and rural construction land shifted back 0.5 km and the center of gravity of Hangzhou stabilized during this period (Fig. 12.3).

12.4   Discussion

12.4.1   Spatial Growth Pattern of Urban and Rural Construction Land

2000–2005: "single-Fan Unbalanced Growth Mode—Enclave Expansion"

From 2000 to 2005, the total amount of urban and rural construction land in Hangzhou increased slowly, and the construction land of urban and rural areas increased in the direction of Xiasha and Binjiang. On the macro perspective, the expansion mode stage was a single fan unbalanced growth mode. The micro land growth mode was the extended enclave type. The growth area was concentrated to the main land of Xiasha in the northeast and southeast of the riverside area and appears to have the enclave extension type. (Fig. 12.2-C1).

2005–2010: "semi-Circular Balanced Growth Mode—Spreading Expansion"

From 2005 to 2010, the total amount of urban and rural construction land in Hangzhou increased rapidly, and the construction land of urban and rural area gradually increased to the two sides of Xiaoshan and Linping and the westside of the city. At the macro level, it was a semi-circular and balanced growth mode, dominated by spreading expansion and supplemented by populating expansion. The main land increment was the spreading expansion in the two sub-cities, Xiaoshan and Linping. In addition, the suburban area of Xingqiao, Dingqiao, Sandun, Kangqiao, Liuxia wre also expanding (Fig. 12.2-C2).

2010–2015: "full Circle Balanced Growth Mode—Form Filling Expansion"

From 2010 to 2015, rapid radial growth of the urban and rural construction land occured in Qianjiang New City construction land, Qianjiang Century City, Dajiangdong, Future science and technology center. It appeared to be whole round balanced growth mode on the macro stage, with filling extension type on micro stage which means that undeveloped land surrounded by existing built-up area was converted into urban construction land. Major spatial growth concentrated in the east and the west with filling extension type. In addition to the West Lake scenic area, Xianghu Lake tourist resort and Xixi wetland area, the urban and rural construction land expanded with discontinuous extension type (Fig. 12.2-C3).

12.4.2 Correlation of Spatial Policy Types and Role Models

According to different functions of space policy, this paper divides spatial policy into three parts: adjustment policy, leading policy and restrictive policy. The adjustment policy adjusts ownership in an area for coordinated regional development, leading policy directs the development of spatial growth that is conducive to social and economic development, and restrictive policy prevents the spread of certain undesirable tendencies to protect the ecological environment (Table 12.2). Adjustment policy is implemented by means of an institutional adjustment, such as the town entrance area, the withdrawal of the city construction area, and the establishment of

Table 12.2 Space policy category and space policy specific action mode

Space policy category	Adjustment policy	Leading policy	Restrictive policy
Space policy Action mode	• Merge the town into the area • Remove "city" to establish "district" • Set up agglomeration area	• Establish industrial platform • Set up a high education zone • Set up a tourist resort • New centres • Transport infrastructure	• Ecological function area protection • Three four lines • Urban development boundary

an agglomeration area. The leading policy affects development mainly through the establishment of industrial zones, high education zones, resorts, new urban centers, and other reconfiguration industries that guide urban function upgrading. Restrictive policy mainly involves the protection of ecological space and games with the adjustment policy and guiding policy to achieve the conjugate balance of ecological and construction space.

Optimized the Urban Structure Under the Action of Adjustment Policy

The early stage of regulatory policy begins with adjustment policy by change the administration from countryside to city system. Later market interventions like setting up industrial agglomeration zones bring fine-tuning in the urban spatial structure. In 1993, Xiasha was adjusted from the Yuhang area into the Jianggan district and set up as an economic and technological development zone in 1996. Three towns from the Xiaoshan area were set up as the national high-tech development zone (Binjiang). From 2000 to 2005, the spatial growth of Xiasha and Binjiang changed the original development mode of the city to the north, and then increased to the southeast of the Binjiang and the eastern part of the city. At the same time, the structural center of urban development shifted eastward. In 2001, the cities of Xiaoshan and Yuhang adjusted to districts, providing new space for urban development, which led to urban development structure changes in the northeast (Table 12.3).

City Function Upgrade Under the Action of Leading Policy

To help enhance contact between the main city and suburb area, the guiding policy transfers industries from the main city to suburb area. At the same time, the guiding policy proposes the construction of transportation infrastructure to strengthen the connection between the suburbs and the main city. At the beginning of the twentieth century, Hangzhou Steel, Hangzhou Oxygen and other industrial companies moved out of the main city to the outskirts. "Traffic to the West" highway construction developed westward while the Hangzhou East Station was developed. At the same

Table 12.3 Hangzhou spatial policy and the correlation analysis of urban and rural construction land evolution model

2000–2005	Directions	N	NEE	SEE	SSW	W
	District	Banshan	Xiasha	Binjiang	Zhijiang	Sandun
	Policy type	Leading policy	Adjustment policy/leading policy	Adjustment policy/leading policy	Leading policy	Adjustment policy/leading policy
	Spatial police	Main city evacuation, industrial relocation	Adjust the town to the district, establish the industrial zone	Adjust three towns, set up high - tech zone	Set up the tourism resort	Adjust the town into the West lake district
	Mode	The spread type	The enclave type	The enclave type	The enclave type	The spread type
	Evolution graphic					
2005–2010	Directions	NE	NEE	E	SW	NWW
	District	Linping	Xingqiao/Dingqiao	Xiaoshan	Liuxia	Sandun/Kangqiao
	Policy type	Adjustment policy/Leading policy	Leading policy	Adjustment policy/Leading policy	Leading policy	Leading policy
	Spatial Police	Remove the city construction area, set up Qianjiang economic and technological development zone	Industrial relocation of main city	Remove the city construction area, set up Xiaoshan economic and technological development zone	Traffic westward, set up education zone	Lead residential area head to west
	Mode	The spread type	The filling type	The filling type	The filling type / The point and axis type	The spread type

(continued)

Table 12.3 (continued)

Evolution Graphic					
2010–2015	Directions	SEE	E	W/NWW	SSW	SWW
	District	Qianjiang New Centre /Qianjiang Shiji Centre	Dajiangdong	Future science and technology centre	Zhijiang	West Lake
	Policy type	Leading policy	Adjustment policy/ Leading policy	Adjustment policy/ Leading policy	Leading policy	Reverse policy
	Spatial police	Catalytic big event and effect, city government move	Construction industrial agglomeration area	Build a great corridor to the west of the city	Yunqi charactistic town and other high-tech industrial zone	West lake scenic spot protection
	Mode	The filling type	The filling type	The filling type / The spread type	The filling type / The spread type	The discrete type
	Evolution Graphic					

time, Yuhang High-Speed construction promoted the spread extension. It appeared that better infrastructure facilities strengthen the connection of urban and suburban, providing support for outside city expansion. The development of technology industries had greatly promoted the implementation of scientific research incubation sites, such as the Zhejiang University development zone. The construction of the tourism resort and the protection development of Xixi wetland had contributed to the prosperity of the urban tourism leisure industry. The new town center of Qianjiang aimed to build the financial harbor and promote the development of the financial service industry headquarters. The west Kechuang corridor gathered the Alibaba net group and other Internet and industry clusters that has deeply improved the creative industry in Hangzhou (Table 12.3).

Balancing Urban and External Space Under the Action of Reverse Policy

Positive policies refer to the adjustment policy and leading policy that promote the growth of urban space, while reverse policies refer to restrictive policy that curbs the spread of spatial growth. Restrictive policy mainly includes single factor protection and space control. The former always divides area and protects water, farmland and forest. The latter establishes ecological function reserves, including ecological function control areas in urban planning space control, drawing a clear distinction between the optimum building area, building area, and the forbidden building area. In 2007, Hangzhou forestry proposed to construct six ecological belts, six country parks and five wetland parks for single factor protection. Hangzhou city master plan (2007 edition) in the space control partition figured to make a clear division of the three areas. There was a game between the restrictive negative policy, the leading, and the adjustment of the positive policy. The negative policy of Hangzhou played a certain role from 2010–2015, but with far less effect than the positive policy. Urban sprawl was caused by the policy transition and negative policy failure in Hangzhou space policy.

12.4.3 Mechanism of Spatial Policy

The Shape Forms Under Spatial Policy Influence

Different spatial policy types have different influences on urban spatial morphological changes, which are characterized by the growth of flying land, spreading type, filling type growth, point-axis growth, and discontinuous growth. The effect of adjustment policy on the urban space form in early years is not obvious due to the lack of spatial entities. When combined with the leading policy, it will soon bring obvious construction land extension. Under the leading policy, the urban and rural construction land form is extended from the main city to the suburbs in the early stage, and the middle ground of the main city and suburbs are soon filled with development. At the same time, the leading policy also promotes regional connection through transportation infrastructure, and promotes the axial expansion of the land form point. Under restrictive policy, urban space skips the control area to show discontinuous growth during the growth process.

The Timeliness Differs Under Spatial Policy Influence

The space policy has different effects on the construction land of urban and rural areas, which is mainly effective, hysteresis effective, and ineffective. For example, leading policy with immediate effect that sets up the industry zone and deploys infrastructure can achieve maximum utility quickly. The adjustment policy has the lag effect, and is often in the form of major policy, so the influence period is long. Restrictive policy often appears to be ineffective, because the protection policy of ecological elements like water, field and forest overlap and conflict. Negative Restrictive policy often fails because the departmental system is not clear due to the transition of positive growth policy.

The Stage Changes Under Spatial Policy Influence

In different stages of development, different types of space policy have difference on the role of urban development. The first period of urban development belongs to urban structural stage, and the urban structure is reorganized under the effect of the adjustment policy. The second period belongs to urban development stage, it appears that leading policy will optimize and upgrade urban functions. The final period of urban development belongs to stable stage, adjustment policy and leading policy combine with negative restrictive policy to balance incremental development with ecological space.

12.5 Conclusion

During the transformation of urban planning from material space planning to public policy management, space policy became an important means for government to transform and improve spatial order. Urban space policy can influence urban structure, urban form, and city function extensively. According to the connotation of space policy, it can be divided into different types: adjustment policy, leading policy and restrictive policy. The space policy and urban spatial growth mode correlation analysis flows that: (1) city structure optimization under the action of adjustment policy; (2) city function upgrade under the leading policy; (3)balance urban and external space under the action of reverse policy. Finally, it reveals that the influence mechanism of space policy on urban and rural construction land is morphological, timeliness and stage. The results and discovery will play an important role in the effective implementation of urban spatial growth management. From 2000 to 2015, leading policy process "hot spots" from center city to suburbs. We can see that the expanding policy is excessive, but restrictive policy is deficient, which becomes to be a problem for big cities like Hangzhou. In the next stage, urban development should abandon incremental growth with no limits.

References

Chang, X., Haijun, W., & Anqi, Z. (2017). Urban spatial multi-scenario simulation and policy analysis of coupling control effect. *Human Geography, 3,* 68–76.

Chenfei, J., & Qun, W. (2015). Study on the policy efficiency evaluation of urban land intensive use based on policy quantification–taking nanjing city as an example. *Resource Science, 37*(11), 2193–2201.

Cheng, L. (2015). Legislation of urban planning public policy from the perspective of legal boundary: Enlightenment from the evolution of planning laws. *China City Planning Review, 24*(02), 47–55.

Cheng, L., Lingyun, F. (2008). On the public policy status and restriction of urban planning. *Planner, 24*(1), 87–90.

Congjian, L., Jingfeng, H., Li, S., & Heyuan, Y. (2013). Study on urban development of hangzhou based on remote sensing. *Urban Development Research, 20*(06), 58–63.

Dennis Wei, Y. H., Liefner, I. (2011). Globalization, industrial restructuring, and regional development in China. *Applied Geography, 32*(1).

Dikeç, M. (2002). Police, politics, and the right to the city. *Geo Journal, 58*(2–3), 91–98.

Doern, G. B., & Pal, L. A. (1988). Public policy analysis: An introduction. *Canadian Public Policy, 14*(2), 225.

Jian, F., & Yu, L. (2008). Outlook on public policy of urban planning in China [J]. *Urban Planning, 4,* 33–40.

Jiansheng, W., Jie, F., & Yang, G. (2014). Study on the ecological effect of urban land policy based on DLS model–taking shenzhen as an example. *Journal of Geography, 69*(11), 1673–1682.

Jingxiang, Z., Tao, Y., & Xiaoling, L. (2013). Urban big event marketing effect in the era of globalization: Based on spatial production perspective. *Human Geography, 28*(05), 1–5.

Li, L., & Yilun, X. (2017). Evaluation of spatial policy based on policy measurement–taking nanjing as an example. *Urban Problem, 6,* 96–103.

Liang, W. (2016) Analysis of the expansion and spatial morphological evolution of urban and rural construction land in Beijing. *Urban Planning, 40*(01):50–59+65.

Lihua, Xu., Huanhuan, W., Jiecun, Z., Wenze, Y. (2014). Spatial and temporal evolution of land use structure in hangzhou in the past 15 years. *Economic Geography, 34*(07):135–142.

Ming, Z., & Chen, C. (2013). The realistic situation, theoretical interpretation and policy thinking of urbanization in China. *Urban Planning, 37*(12), 9–21.

Mingyue, N. (2012). China's urbanization characteristics, problems and governance. *Nanjing Social Science* (10), 19–27.

Salet, W., & Thornley, A. (2007). City-regions institutional influences on the integration of multilevel governance and spatial policy in European. *Journal of Planning Education & Research, 27*(2), 188–198.

Shiying, W., Daming, C., Weifeng, K. (2017) Research on the implementation of security system for the new urban general planning in Shanghai based on public policy. *Journal of Urban Planning,* (s1).

Song, S. (2007). Urban spatial organization–an analytical framework for urban policy and planning. *The Planner, 23*(11), 28 30.

Wang, H., He, Q., Liu, X., et al. (2012). Global urbanization research from 1991 to 2009: A systematic research review. *Landscape & Urban Planning, 104*(3–4), 299–309.

Wei, W., Min, C., & Nan, W. (2017). Advances in the study of spatiotemporal heterogeneity in urban land use in China. *Geography and Geographic Information Science, 6*, 57–63.

Wenyu, W. (2014). *Study on the dynamic mechanism of urbanization in China from the perspective of government behavior*. East China normal university.

Wenze, Y., Ruiliang, W., & Beilei, F. (2013). Study on the spatial pattern of urban expansion–taking hangzhou as an example. *Journal of Zhejiang University (science Edition), 40*(05), 596–605.

Yao, S., Zhang, P., & Yu, C. (2014). The theory and practice of new urbanization in China. *Scientia Geographica Sinica, 34*(6), 641–647.

Yonghong L. (2005). On the importance of space policy--taking Shenzhen as an example. *Urban Planning, 29*(12):45–53. 42.

Chapter 13
Study on China's Adaptation Policies to Climate Change from the View of Urban and Rural Planning

Yuan Huang and Yanxiao Pan

13.1 Introduction

In recent years, with rapid urbanization, the urban heat island effect has become more and more serious, and extreme weather disasters such as urban floods and droughts have increased in frequency. Cities in China are facing dual challenges: maintaining economic growth and responding to climate change. Regarding urban and rural responses to climate change, Yamaguchi et al. have indicated that effective design, planning, building layouts, and energy consumption equipment systems in urban blocks could reduce CO_2 emissions by 60–90% of current emissions by the middle of the twenty-first century (Yamaguchi et al., 2007). The UN Habitat has initiated a project titled the "Cities and Climate Change Initiative". The project's Assessment Report states that properly planned cities consume spaces and energy in a more efficient way by reducing the traffic flow and infrastructure construction, thus reducing the impact of climate change (Cities and Climate Change Initiative, 2014). Rational planning in urban and rural areas has become the "main force" to climate change response.

In order to better respond to climate change, Chinese government has successively released a series of relevant policies ranging from "mitigation" to "adaptation".[1] Within policy development, urban and rural planning has played a key role and gradually become the focus for adaptation polices, (Li et al., 2011) with a goal

---

[1] According to *China's national climate change program (2007)*, climate change mitigation and adaptation are two organic components of the challenge in climate change response. They are abbreviated as "mitigation" and "adaptation" in the following paragraphs.

Y. Huang (✉) · Y. Pan
Southwest Jiaotong University, Xibu Park, Chengdu, Sichuan, China
e-mail: yuanhuang@home.swjtu.edu.cn

Y. Pan
e-mail: YanxiaoPansue@163.com

© Springer Nature Switzerland AG 2021
W. Li et al. (eds.), *Human-Centered Urban Planning and Design in China: Volume I*,
GeoJournal Library 129, https://doi.org/10.1007/978-3-030-83856-0_13

of integrating relevant climate change response indicators into urban and rural planning systems, construction standards, and industrial development plans through the year 2020 evidenced in the *"Urban Adaptation to Climate Change Action Plan (2014–2020)"* (Development and Reform Commission (NDRC), 2017). Therefore, the analysis of adaptation policies from the view of urban and rural planning mutually promotes the development modes of "urban and rural planning" and "accommodating climate change" in China.

13.2 Urban and Rural Planning Under the Background of China's Mitigation Policies

From the development of China's mitigation policies, (Development and Reform Commission (NDRC), 2007, 2010, 2014) the Chinese government has paid more and more attention to the development of urban and rural planning for mitigation. The focuses are formulated from simple green building technologies to energy conversation and emission reduction in various areas of urban and rural planning. Low-carbon emission and green planning have been implanted in the aspects of land use, buildings and transportation in urban and rural planning (Table 13.1). The low-carbon emission province-city/town-park /community and green ecological city planning have promoted green transportation and buildings at larger scales. A number of demonstration projects have been carried out in rapid succession at different scales, such as green and low-energy buildings, low-carbon emission transportation, low-carbon emission provinces-cities/towns-parks/communities/ and green eco-cities (Table 13.2).

On the whole, urban and rural planning in China's mitigation policies aims at energy conservation and emission reduction and focuses on low-carbon emission and green planning. The planning employs quantitative indicators for "carbon emission" control and has carried out a number of demonstration construction projects as promotional measures, which presently are achieving great success.

13.3 Urban and Rural Planning in China's Adaptation Policies: Analysis at the National Level

13.3.1 Definition of Adaptation in Urban and Rural Planning

Compared to the clear aim and definition of climate change mitigation, there has been no clear and unified definition and explanation on climate change adaptation in the world. The default assessment of climate change mitigation is the quantitative indicator on CO_2 emissions in the international community. However, the understanding of the concept of adaptation changes within various contexts of local and regional culture, politics and economies. Presently, China has not given a clear explanation of

Table 13.1 Urban and rural planning in China's mitigation policies (Development and Reform Commission (NDRC), 2007, 2014)

Building	Transportation	Land	Industry
1.Green building Construction area proportion, standard execution value and execution rate **2. Energy-saving building** Standard execution value and execution rate **3. Green building material 4. Existing building energy conservation transformation** Different climate zones and different types of buildings **5. Renewable energy building** Solar, geothermal and other renewable energy building proportions	**1. Vehicle control 2. New energy vehicle 3. Transit-oriented development (TOD) 4. Green transportation** Public, slow and intelligent transportation **5. The proportion of public transportation in the total transportation types**	**1. Urban functional layout** Work-life balance, group layout, mixed development, compact city **2. Smart growth and scale control**	**1. Adjust economic structure, upgrade technology standards 2. The integration of industries and residences**

adaptation which may lead to significant ambiguities and limitations on the definition of adaptation in urban and rural planning.

The definition of adaptation promoted by the Intergovernmental Panel on Climate Change (IPCC, 2007) is "adjustments in natural or human systems in response to actual or expected climatic stimuli or their effects, which moderate harm or exploit beneficial opportunities" (Intergovernmental Panel on Climate Change (IPCC), 2007). Conceptually, adaptation is increasingly conceived of as the management of climate risk. This connects adaptation to broader perspectives on urban resilience (Carter et al., 2012).

In China, with the acceleration of urbanization, urban land has become extremely scarce. High-density and high-intensity construction modes are a common phenomenon in urban centers, leading to climate change issues such as the heat island effect and hazy weather. At the same time, drought, floods and other extreme weather occur frequently in recent years, bringing great threats to people's health, life and property due to the vulnerability of infrastructure. Therefore, adaptation shold link closely to climate environment improvement and climate risk management in urban and rural planning.

Table 13.2 Demonstration projects on low-carbon emission and green cities, towns and districts (National Development and Reform Commission (NDRC), 2014, 2017; Ministry of Industry and Information Technology, National Development and Reform Commission (NDRC), 2013; Ministry of Housing and Urban–Rural Development (MOHURD), 2013, National Development and Reform Commission (NDRC), Ministry of Housing and Urban-Rural Development) (MOHURD), 2011)

Low-carbon emission provinces-cities (The first batch: 13 pilots in 2010; the second batch: 29 pilots in 2012; the third batch: 45 pilots in 2017)	Low-carbon emission parks (The first batch:55 pilots in 2014;, the second batch: 80 pilots in 2020)	Low-carbon emission communities (1000 pilots until 2015)	Low-carbon emission towns (The first batch: 7 pilots in 2012; 30 pilots until 2020)	Green ecological demonstration districts (The first batch in 2010;100 pilots until 2015)
1. Low-carbon emission development planning 2. Low-carbon emission and green development supporting policies 3. Low-carbon emission industrial system 4. Low-carbon emission green lifestyle and consumption mode	**1. Low carbon emission production design** **2. Low carbon emission technology research, development, application and industrialized development** **3. Carbon emission management system** **4. Low-carbon emission infrastructure** **5. International cooperation**	**1. Low carbon emission property management and service new modes** **2. Energy-saving and green building** **3. High efficiency and low carbon emission infrastructure** **4. Low-carbon emission public transportation** **5. Low-carbon emission living facilities** **6. Ecological environment planning and design:** (1) Green belt noise reduction (2) Green walking road (3) Public green space	**1. Low-carbon emission industrial system** Adjust economic structure **2. Low-carbon emission lifestyle** **3. Low carbon emission planning** (1) Optimize infrastructure construction (2) Public transport (3) Energy and water systems (4) High quality construction	**1. Land-use** (1) Mixed development (2) Reasonable road network density (3) Public service convenience **2. Green building** (1) Green energy (2) Green materials **3. Green transportation** **4. Industry and economy** (1) City and industry integration (Work-life balance) (2) Industrial structure optimization (high-tech industry and green industry)

13.3.2 Main Areas of Urban and Rural Planning in Adaptation Policies

The development of China's adaptation policies (Development and Reform Commission (NDRC), 2007, 2013, 2014, 2017) has ranged from "zero" to "one", from the "whole" to the "specialized", and from simply focusing on urban infrastructure construction to improving buildings' adaptability, promoting ecological greening, protecting infrastructure safety and completing disaster risk management, all of which have realized the exploratory development of multiple areas in urban and rural planning (Table 13.3). The main areas involved in urban and rural planning in China's adaptation policies to climate change are as follows:

Table 13.3 shows that urban and rural planning in China's adaptation policies is still in an initial exploratory stage. The features of the policies can be summarized as the:

(1) Integration of "hard and grey" and "soft and green" planning
The polices adopt the integration mode of "hard and grey" measures and "soft and green" measures in urban and rural planning. The "hard or grey" planning focuses on reducing disaster risk and relieving the vulnerability of buildings and infrastructures. The "soft" or "green" planning aims at ameliorating microclimates, conserving soil and water, etc., through ecological greening and river water system planning (Wamsler et al., 2013) which has become an important development trend in relevant areas of urban and rural planning for adaptation in recent years with the development of ecological civilization construction.

(2) Unbalanced qualitative and quantitative descriptions
Most of the planning measures related to the adaptation policies are qualitative descriptions except for the aspect of ecological greening. A complete quantitative assessment system for adaptation has not yet been developed.

(3) Combination of multiple areas in urban and rural planning

The relevant areas in urban and rural planning for adaptation are diversified, including resilient urban planning, sponge city planning, water-saving city planning, flood prevention and drainage planning, ecological greening planning, building adaptability planning, comprehensive disaster prevention demonstration community planning and underground space planning. The important connotation of "adaptation" is contained in these specialized plans which insert adaptation actions into various areas of urban and rural construction.

Table 13.3 Urban and rural planning in China's adaptation policies to climate change (Development and Reform Commission (NDRC), 2013, 2017)

Urban and rural planning guidance	Infrastructure adaptability	Buildings' adaptability	Ecological greening	Urban water security	Comprehensive Urban disaster risk management system
1. Climate change factors are considered in urban planning Urban location selection, urban expansion, urban planning, etc (1) Climate change risk assessment and climate feasibility study (2) Reasonable control of population and land use scale based on the resource carrying capacity **2. Strengthen the layout of related fields**	**1. Improve the standards for urban infrastructure construction** **2. Disaster prevention and reduction planning, underground space planning, and vertical planning** **3. Sponge city, water-saving city, and a comprehensive disaster prevention demonstration community** **4. Resilient city and flexible city planning**	**1. Urban renewal and old residential comprehensive reconstruction** **2. Industrialized promotion of prefabricated buildings** **3. Improve the ventilation design of buildings, advocate for roof greening and three-dimensional greening**	**1. Establish a climate-friendly urban ecosystem** Open up city ventilation corridors, etc **2. Improve urban microclimate through landscape** (1) Create green space, wetland, town square and other diversified green spaces (2) Control urban park green area, urban green land rate, green coverage rate, green land rate in urban residential areas and other indicators **3. Urban ecological space protection and control** (1) Urban main functional area planning (2) Ecological restoration projects **4. Ecological city, garden city planning**	**1. Sponge city** **2. Water-saving cities** **3. Urban flood control system** **4. Reasonable planning of river systems**	**1. Urban emergency services** **2. Public alert protection system** **3. Risk sharing mechanism**

13.4 Urban and Rural Planning in China's Adaptation Policies: Analysis at the Regional Level

13.4.1 Overview of the Pilot Cities' Development

The pilot cities' development have been divided into three types of adaptation zones in the east, central and west regions in China based on geographical location and climate features. And there are 28 pilot cities with different climate risks, scales and functions that have been selected as the climate change adaptation pilot areas (National Development and Reform Commission (NDRC), 2017). Presently, the preliminary adaptation plans have been gradually released and a series of adaptation actions have been carried out in accordance with regional conditions in these pilot cities.

13.4.2 Analysis of Urban and Rural Planning in the Representative Adaptation Pilot Cities

In this paper, three pilot cities have been selected among 28 pilot cities: Chaoyang City of Liaoning Province, Qiannan District of Chongqing City and Baise City of Guangxi Province, representing the eastern, central and western zones, respectively. This paper explores the main problems within the development of the pilot cities and proposes solutions, strategies and promotional suggestions for urban and rural planning for adaptation.

The analysis of the main areas of urban and rural planning in the representative pilot cities are listed as follows (Table 13.4).

Based on the above analysis, the features of urban and rural planning in adaptation policies among different pilot cities can be summarized as the:

(1) Development of localization of urban and rural planning
   Most of the pilot cities have similarities in urban and rural planning for adaptation policies. For example, they tend to focus on the optimization of urban infrastructure layout and construction standards, the promotion of urban renewal, the comprehensive transformation of old residential communities, the promotion of industrialization of fabricated buildings, the optimization of temperature regulation functions and land and water maintenance in river networks, and the improvement of the flood control and drainage system construction such as in sponge cities (National Development and Reform Commission (NDRC), 2017). However, in few cities they have taken into account in their adaptation policies large-scale disaster risk.
(2) Comparison of the planning measures in developing and developed cities

Table 13.4 Urban and rural planning in the representative pilot cities for adaptation (Chaoyang Government, 2017; Chongqing Longnan District Government, 2017; Baise Government, 2018)

Chaoyang City of Liaoning Province	Qiannan District of Chongqing City	Baise City of Guangxi Province
Climate change impact and vulnerability assessment **Action Plans for adaptation** 1. Scientific planning of urban lifeline systems (a) Climate factors are considered in planning (b) Strengthen the layout of related fields **Organize actions to adapt to climate change** 1. Improve the adaptability of buildings (a) Strict implementation of building energy efficiency standards (b) Implementation of urban renewal and comprehensive renovation of old residential areas (c) Green ecological city construction 2. Develop the function of ecological greening (a) Building a Climate-friendly Urban Ecosystem (b) Develop landscape to improve the microclimate 3. Ensure urban water security (a) Promoting the construction of a sponge city (b) Build a comprehensive water-saving city (c) Reasonable urban flood control system 4. Build a green traffic demonstration city (a) Give priority to the public transport 5. Underground space utilization and management **Adaptation capacity construction**	**Strengthen urban planning guidance** 1. Planning should play a coordinating role in the climate carrying capacity 2. Increase the countermeasures for specialized planning about climate capacity 3. Optimize the functional layout of urban space **Establish a traffic network for adaptation** 1. Smooth traffic network **Rational allocation of water resources for adapting to climate change** **Improve the urban functional supporting facilities for adaptation** 1. Configure city requirements (a) Construction of forest park and wetland park 2. Strengthen urban management (a) Old City Reconstruction (b) Facilities for flood control, drainage, water, electricity, and garbage disposal 3. Improve underground pipe network 4. Promote smart city construction 5. Sponge city planning **Improve the quality of buildings that adapt to climate change** 1. Accelerate the construction of green buildings 2. Focus on cultivating the green building materials industry **Create an urban ecological environment that adapts to climate change** 1. Dress up the urban landscape 2. Strengthen environmental governance 3. Develop ecotourism 4. Strengthen ecological restoration 5. National forest city **Strengthen the urban capacity to deal with climate change**	**Popularize the idea of urban adaptation** 1. Integrating adaptation with urban relevant planning 2. Improve the planning and layout of public service facilities in the downtown area **Improve monitoring and warning capabilities** **Carry out specialized actions in the key areas** 1. Improve infrastructure in the downtown area 2. Promote green buildings and carry out transformation of old districts 3. Strengthen the ecological construction of the central city 4. Ensure urban water security (a) Sponge city construction in the central areas (b) Underground comprehensive pipeline construction in central areas (c) Optimize flood prevention and drainage systems in the central areas (d) Strengthen water resources protection and management **Strengthen capacity for climate change response**

Most of the developing cities stayed in the level of "hard or grey" planning. However, the relatively developed cities began to pay more attention to the "soft or green" planning, such as afforestation and river systems planning.

(3)   Analysis of the role of water-related planning in adaptaion policies

Water security has become a common denominator in different pilot cities. Water safety planning and management generally appear as separate clauses in different action plans. Moreover, a series of water-related planning has been carried out, such as resilient urban planning, sponge city planning, water-saving urban planning, flood prevention and drainage planning, and ecological greening planning.

13.5   Discussion

From the analysis above, it can be found that urban and rural planning in China's adaptation policies is still in the initial exploratory stage, due to the lack of understanding on adaptation, the weak foundations and the incomplete systems, etc. (Development and Reform Commission (NDRC), 2013, 2017). The potential to improve urban and rural planning in the China's adaptation policies is discussed in the following aspects:

(1)   **Geographic areas**: strengthening rural and old urban areas' planning.
It can be found that the main emphasis in urban and rural planning is on the developed urban areas and new urban areas in China. However, in the vast rural and old urban areas, the planning for adaptation has just begun and is moving at a slow pace. Supported by of "rural rejuvenation" and "urban renewal" by the central government, we should gradually develop planning in rural and old urban areas. In rural areas, it can improve the capacity to deal with climate risks. In old urban areas, it can improve the capacity to adapt to climate change through renovation and reconstruction.

(2)   **Construction areas**: further strengthening water security planning and building adaptability planning.
In adaptation policies, water system planning and management are particularly important. Rational water safety planning and management can reduce various climate change risks such as floods, landslides, extreme temperatures, urban droughts, and the urban heat island effect (Wamsler et al., 2013), linking closely with urban ecological greening, urban river network planning, etc. Therefore, the *First Line of Life Security: Water Security* should be further strengthened in urban and rural planning. In current urban and rural planning, the consideration for adaptive building construction is inadequate; most of building planning is still focused on the construction of green buildings which is more closely related to mitigation rather than adaptation. So adaptation policies should strengthen the researches of the deeper connotation of building adaptability.

(3) **Morphological design**: strengthening the design of architectural, vegetation, surface coverage morphology, etc.

Current urban and rural planning is lacking relevant morphology design. However, the layout of urban morphology such as architectural layout patterns, architectural density, architectural height, architectural orientation and vegetation space types have a greater decisive effect on the local climate in cities. Therefore, the design of cities creates unique micro-climates that affect variables including temperature and wind (Keskitalo et al., 2012) Adaptation policies should strengthen the research on the influence mechanisms between urban morphology and urban thermal environments in urban and rural planning, so as to strengthen the rational design of urban morphology.

(4) **Development of localization**: strengthening the consideration of local climate background and terrain conditions.

Due to the lack of basic data for the prediction, assessment and feasibility demonstration of potential risks of climate change, urban and rural planning in most cities lacks sufficient consideration of specific local contexts. Adaptation is closely related to local conditions as the locus of control and benefits associated with adaptation resides locally (Carter et al., 2015). There is an obvious local feature in adaptation action. Therefore, urban and rural planning should consider local climate background and terrain conditions more systematically so as to carry out more effective and customized measures.

13.6 Conclusion

This paper analyzes China's national and regional adaptation policies from the view of urban and rural planning. The above analysis finds that the importance of the role of urban and rural planning in the adaptation policies is increasing day by day. There is a great potential for urban and rural planning in promoting the development of adaptation polices, which can be prospected from the following aspects:

(1) **Urban and rural planning is conducive to the realization of localization development of adaptation**. There is an obvious local feature both in urban and rural planning and adaptation actions. Through the continuous improvement of urban and rural planning, which is controlled by local government, the adaptation actions can be implemented more effectively and specifically.

(2) **Urban and rural planning is conducive to the realization of multiple areas of development of adaptation**. The urban and rural planning decision-making needs to comprehensively consider the specialized plans of meteorology, municipal administration, transportation, and disaster prevention, all of which are closely related to adaptation. Therefore, improving the status of urban and rural planning within adaptation policies can promote the integration of multi-sectoral planning.

(3)   **Urban and rural planning is conducive to the realization of sustainable development goals of adaptation**. The ultimate goal of adaptation is to achieve sustainable development. Paying attention to urban security and equity issues, such as distributing urban resources equitably through public transport, housing, environmental, and social security policies will help achieve this goal (Zheng, 2012). As an important means of comprehensive allocation of urban and rural resource space, urban and rural planning is aligned with adaptation actions with the goal of sustainable development.

Adaptation actions have only just begun. Therefore, in a follow-up study for this paper, we will continue tracking the development of the 28 adaptation pilot cities, and strive to explore the development mode of urban and rural planning in China's adaptation policies more deeply. We will further exploit the potential for the application of urban and rural planning in adaptation with the hope to integrate adaptation into various systems of urban and rural planning.

Acknowledgements   The authors would like to thank the National Natural Science Foundation Project of China (grant number: 51508496) for supporting this study.

References

Baise Government. (2018). *National Climate Adaptation City Construction Pilot Action Plan of Baise*. [EB/OL]. http://www.baise.gov.cn/www/zww/html/2018-02/201802221048193884.html.

Carter, J. G., Cavan, G., & Connelly, A. (2015). Climate change and the city: Building capacity for urban adaptation. *Progress in Planning., 95*, 1–66.

Carter, J. G., Connelly, A., Handley, J., & Lindley, S. (2012). *European cities in a changing climate: Exploring climate change hazards, impacts and vulnerabilities*. The University of Manchester.

Chaoyang Government. (2017). *Climate-Adaptive City Construction Pilot Work Plan of Chaoyang*.[EB/OL]. http://www.zgcy.gov.cn/ZGCY/zwgk/20170920/004006001_5c4b84c1-7eb4-46d2-b64c-728ef2da79ff.htm.

Chongqing Longnan District Government. (2017). *Climate Adaptable City Pilot Implementation Plan of Weinan District*. [EB/OL]. http://tn.cq.gov.cn/xxgk/nēws/2017-12/19_53202.shtml.

Cities and Climate Change Initiative. (2014) *Tool series, planning for climate change: a strategic, values-based approach for urban planners*. United Nations Human Settlements Programme, UN-Habitat.

Intergovernmental Panel on Climate Change (IPCC). (2007). Climate change. (2007). *AR4 synthesis report: Contribution of working groups I, II and III to the fourth assessment report of the Intergovernmental Panel on Climate Change[R]*. Cambridge University Press.

Keskitalo, E., Georgi, B., Isoard, S., et al. (2012). *Urban adaptation to climate change in Europe: Challenges and opportunities for cities together with supportive national and European policies(R)*. European Environment Agency, (EEA).

Li, Y., Yang, X., & Zhu, X. (2011). Integrating climate change factors into China's development policy: Adaptation strategies and mitigation to environmental change. *Ecological Complexity, 8*, 294–298.

Ministry of Housing and Urban-Rural Development) (MOHURD). (2013). Notification of "Twelfth Five-Year" development planning on green building and green eco city. [EB/OL]. http://www. gov.cn/gzdt/2013-04/18/content_2380994.htm.

Ministry of Industry and Information Technology, National Development and Reform Commission(NDRC). (2013). *Notification of pilot work on national low carbon industrial park.*[EB/OL]. http://www.miit.gov.cn/n1146295/n1652858/n1652930/n3757016/c3762117/content.html.

National Development and Reform Commission(NDRC), Ministry of Housing and Urban-Rural Development)(MOHURD). (2011). *Notification of pilot demonstration work on the first batch of green and low carbon key small town.* [EB/OL]. http://www.gov.cn/zwgk/2011-10/28/content_1 980272.htm.

National Development and Reform Commission(NDRC). (2013). National Adaptation to Climate Change Strategy (2013~2020) [EB/OL]. http://qhs.ndrc.gov.cn/syqhbh/201312/t20131213_570 418.html.

National Development and Reform Commission(NDRC). (2017). Ministry of Housing and Urban-Rural Development)(MOHURD). Notification of pilot work on Climate-Adaptive City Construction (2017) [EB/OL]. http://qhs.ndrc.gov.cn/syqhbh/201702/t20170224_839166.html.

National Development and Reform Commission(NDRC). (2017). Urban Adaptation to Climate Change Action Plan(2014~2020)[EB/OL]. http://qhs.ndrc.gov.cn/syqhbh/201602/t20160216_ 792843.html.

National Development and Reform Commission(NDRC). (2007). *China's National Climate Change Program* [EB/OL].http://www.ccchina.gov.cn/Detail.aspx?newsId=28013.

National Development and Reform Commission(NDRC). (2014). National Climate Change Plan (2014~2020) [EB/OL].http://qhs.ndrc.gov.cn/zcfg/201411/t20141105_647420.html.

National Development and Reform Commission (NDRC). (2010). *Notification of pilot work on low-carbon provinces and low-carbon cities* [EB/OL]. http://www.ndrc.gov.cn/zcfb/zcfbtz/201 008/t20100810_365264.html.

National Development and Reform Commission(NDRC). (2014). Notification of pilot work on low-carboncommunity[EB/OL].http://www.ndrc.gov.cn/zcfb/zcfbtz/201403/t20140327_604 483.html.

Wamsler, C., Brink, E., & Rivera, C. (2013). Planning for climate change in urban areas: From theory to practice[J]. *Journal of Cleaner Production, 50,* 68–81.

Yamaguchi, Y., Shimoda, Y., & Mizuno, M. (2007). Transition to a sustainable urban energy system from a long-term perspective: Case study in a Japanese business district. *Energy and Buildings, 39,* 1–12.

Zheng, Y. (2012). Resilient city: Mainstreaming climate risk management and adaptation to climate change into urban planning. *Urban Planning, 01*(19), 47–51.

Chapter 14
Considerations on Urban–Rural Relationship and Planning Philosophy from the Perspective of Rural Planning in Contemporary China–Discussion on the Existing Problems in Rural Planning Education

Yang Fan, Zhou Tianyang, and Zhu Jiehao

14.1 Introduction and Background

"Agriculture, rural areas and villagers" is an important issue of the overall social development with Chinese culture. As an important part of the Chinese national composition, the villagers engaged in the industry, the region where they live and the identity characteristics when they involved in social and economic activities have the complexity of specific historical and environmental conditions, which leads to the formation of dual social system between urban and rural areas in China. Agriculture-related policies always play a dominant role in China's national policies. Following the *New Socialist Countryside Policy* proposed in the Sixth Plenary Session of the 15th Central Committee of the Communist Party of China (2005) and *Beautiful Countryside Policy* proposed in the 18th National Congress of the Communist Party of China (2012), the Chinese Communist Party proposed the *Rural Revitalization Strategy* in the reports of the 19th National Congress of the Communist Party of China.

After the introduction of successive agriculture-related policies, urban–rural planning is considered an important approach which bears the responsibility of carrying

Notes This research project is funded by National Natural Science Foundation of China (Approval Number: 51778436)

Y. Fan (✉) · Z. Tianyang · Z. Jiehao
Urban Planning Department, College of Architecture and Urban Planning, Tongji University, Shanghai, China
e-mail: fanyangsh@tongji.edu.cn

Z. Jiehao
e-mail: zhujiehao@126.com

W. Li et al. (eds.), *Human-Centered Urban Planning and Design in China: Volume I*, GeoJournal Library 129, https://doi.org/10.1007/978-3-030-83856-0_14

the national policies and promoting the social and economic development in rural areas. Moreover, it is also considered that urban–rural planning plays an important role in practicing the national policies concerning the promotion of urbanization, such as *New-type Urbanization Policy, Innovation and Entrepreneurship Policy*, and *Transformation Development Policy*. As such, the value, preparation system, and problem-solving path of urban–rural planning have an important impact on the formation of a good and ordered urban–rural relationship. At present, the planning forms such as rural planning and village planning for rural areas provide "media" and mechanism for exploring the rural governance mechanism and have become the hot issue in the academic circle. As a reflection to this academic hot point, this paper analyzes the problems and dilemmas encountered by all the plans for rural areas, points out the present rural planning education situation and misunderstandings, and thus provides a reference to understand the contemporary urban–rural relationship and planning method in China.

14.2   Basic Characteristics of China Rural Development——A Short Review

Chinese countryside has long been considered the economic and social bedrock of China, while also reflecting the social and economic problems. This paper concludes the following three important characteristics

First, urban–rural dual structure has long been discussed and has constructed the basic context for research on urban–rural relationship in China. We can interpret this in three dimensions, economic, social identity and production factor, as follows: a. For the economic dimension, the adoption of traditional "price scissors" for urban and rural production factors, the setting of strict "rural–urban dual economy threshold", and the utilization of rural areas to support the construction and development of urban areas bring on the fixation of the urban and rural interest differentiation pattern and finally lead to the rapid enlarging of income gap between urban and rural residents (Binkai & Yifu, 2013).

b. For the dimension of social identity, the long-established "urban–rural dual" HUKOU registration system, employment system, and social insurance system induce a fixation on social identity, and consequently reduce the population mobility between urban and rural areas, so that "villager-workers" have evolved into an institutional arrangement and can hardly enjoy the public services in urban areas or protect their basic individual rights (Guosheng 2009). It becomes a dilemma for village residents to decide whether or not to be an urban resident.

c. Based on aforementioned aspects, the circulation of production factors between urban and rural areas is relatively restricted. Though the rural areas have been revitalized at present, we can still discern clearly that "urban–rural dual structure" is the root cause of the differentiation of orders and uneven distribution of resources between urban and rural areas. The dual structure unnaturally influences the urban–rural harmonious development and two-way population flow and maintains the relatively separated urban and rural orders.

Second, rural land use policy is the sticking point for the use and protection of land and space resources of China, as well as for the sustainable development of urban and rural areas. On one hand, because of the different methods for dividing the ownership of urban and rural lands, the use of land for construction must be based on the transfer of land use right. On the other hand, the current rural land regulation system has resolved the subsistence problem of village residents, but it has brought about the fragmentation of land ownership which is unfavorable for the scale management of rural land (Qing et al., 2008) and led to the separation of rural land ownership rights, contract rights, and management rights. Furthermore, the rights and interests of village residents cannot be guaranteed basically because the property rights of land are ambiguous in the rural land regulation system (Kun, 2011). This may affect the rural land transfer and willingness of village residents to be transformed into urban residents, seriously impact the process of transformation of villager-workers into urban residents, directly or indirectly weaken the independence of village residents as economic subjects, and thus restrict the operation of market economic system in rural areas (Xi, 2011).

Third, villager self-governance, as the important content of the grass-roots democracy political system of China, is the basic method of countryside governance. The extension of administrative management authority and organization into the rural areas will absolutely have certain conflict with the villager self-governance system. For example, the reform of agricultural tax and financial policy. Different perspectives of social management and different understanding of construction management will probably affect the steady state of rural politics in different degrees and may cause the political unbalance. This unbalance may beget the contradiction between the Village Party Branch Committee and Villager Self-Governance Committee (Jianrong, 2010), village elites (Sihong, 2003) and villagers (Linxiu et al., 2005) and disclose the dual identity of "villagers committee", i.e., grass-root agent of national will—the "nerve terminal" of township government and grass-root villager self-governance organization. Because township politics and village governance have difference in right competition and interest demand, it is practically difficult for one decision policy-making body to control the competition between two decision-making subjects of different natures (Weiyi, 2004).

Therefore, this paper holds that the current rural planning faces at least three dilemmas. The first dilemma is that planning behavior is not specified by the law; the second is that system frame causes the loss of entrusting subject; and the third is that the land regulation system affects the land use mode. If no attention is paid and no research is conducted, these dilemmas will seriously affect the effect of rural planning in providing mechanism and operation platform for rural social governance.

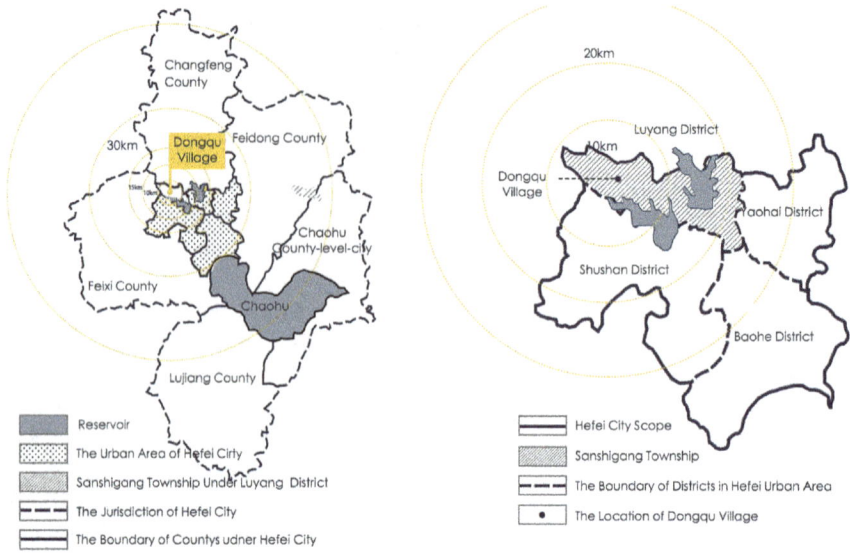

Fig. 14.1 The location of sanshigang township

14.3 Difficulties Encountered by China's Contemporary Rural Planning Based on Case Studies

14.3.1 Introduction to Case Studies

Sanshigang Township[1] is located at the northwest of Luyang District of Hefei and is 22 km away from the downtown of Hefei City. It covers an area of 32.58 km^2. The subject of this study is Dongqu Village, located to the east of Sanshigang Township and 19.9 km away from the downtown of Hefei. The whole village covers an area of 3582.6 mu (2.38 km^2), the current agricultural acreage is 889 mu (0.59 km^2), and the area of farmland converted to forestry is 532 mu (0.35 km^2). There are 8 village groups, covering 531 households and 1679 permanent residents (see Figs. 14.1 and 14.2).

In 2005, Dongqu Village, as one of the new countryside construction demonstration sites of Hefei, conducted overall environment renovation and reconstructed the cottages according to Hui-style and Han Dynasty-style architecture by way of finishing of outer surfaces and integral reinforcement. In 2017, Dongqu Village was selected as one of the main design bases for the First National Rural Planning Competition for College Students.

---

[1] Town and township in Chinese respectively mean urban areas and rural areas under the same administrative level.

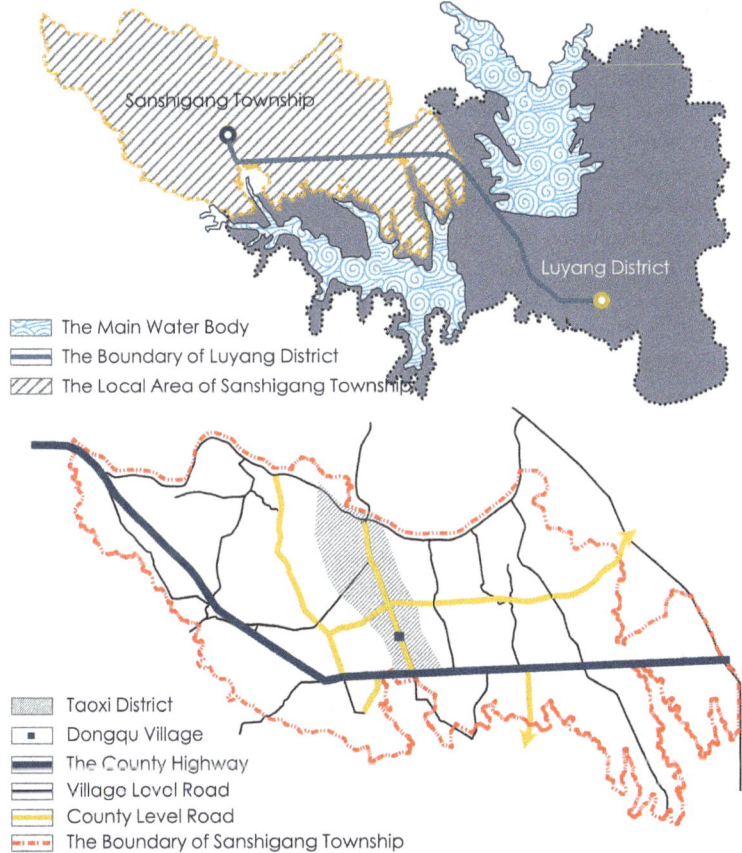

Fig. 14.2 The detailed location of sanshigang township

14.3.2 Insufficient Attention to the Governance System and Legal Background in Current Rural Planning

Planning is a collective vision and management tool of a community for the future living status. It is usually a part of the local administrative body or the governance body power systems and a function of top-down national administration system or an item of the power coordination by the local governance mechanism. There is no exception for rural planning, that it is either the intervention of the government administrative power on rural community or the self-management tool of villager self-governance organization. From this, we can see rural planning is facing the following difficulties:

(1) Rural planning, if it is taken as an external guide and intervention power to rural communities, has an insufficient legal basis. In urban areas, urban planning is the decision-making method of multivariate integrated control which

is dominated by a single decision-making body (Yong & Xiyun, 2005); while in rural society, villager self-governance is a kind of grass-roots democracy organization. Article 111 of the Constitution (2018) clearly specifies that villagers committee is the grass-root mass self-governance organization and that villagers committee sets people's mediation commission, security defense commission, and public health commission to handle public affairs and public welfare undertakings, mediate disputes among the people, assist in maintaining the social security, and reflect the opinions and requirements from the mass to the People's Government. *Organic Law of the Villagers Committees of the People's Republic of China* issued in 1987 also has the same stipulation in Article 2. Therefore, rural planning and village planning should be handled by the villagers committee as one of the public affairs and public welfare undertakings. If the villagers committee does not have the corresponding technology, it should entrust the rural planning to a reliable institution. However, both the current rural planning and village planning carried out in rural areas can be traced back to the *Urban and Rural Planning Law of the People's Republic of China* implemented on January 1, 2008, *Village and Town Planning and Construction Management Regulations* issued on June 29, 1993, and *Famous Historical and Cultural City, Town and Village Protection Regulations* issued on April 22, 2008, which do not regard the preparation and management of rural planning as the autonomous affairs of the villagers committee, but as an extension of the administrative power of the township government. This does not only neglect the basic power of villager self-governance granted by the Constitution, but also weakens the self-governance achievement of the grass-roots democracy and violates the original intention of the constitutional system (Ke & Jing, 2017). Moreover, because the administrative mechanism, laws and regulations are issued in different times and are incomplete in content, the rural construction administration can be shortsighted and planning implementation can become unclear and passive (Jiangang & Hanzi, 2015).

(2) The rural planning implementation and responsibility subject is unclear. The understanding of rural planning in legal sources has led to difficulties in the planning preparation and implementation. The *Organic Law of the Villagers Committees of the People's Republic of China* specifies in Article 5 declares that the people's governments of township, nationality township and town give guidance and help to support the villager committee's works, but may not intervene in the matters which fall within the scope of villager self-governance. The state still chooses to extend its functional power to the village level, even though it grants village residents the political autonomy to conduct self-management. Hence, it is generally the case that the township party and government organizations dominate the village governance (Yiqiang & Jun, 2014). Especially in the rural planning management, due to the imbalance in allocation of economic and social resources, the guidance of the higher-level party and government organizations may easily become a leading relationship, making the villager self-governance meaningless (Qiang, 2012). In fact, opinions tend to be inconsistent in the process of villagers self-governance. When the rural planning and

construction is taken as a certain indicator for assessing the achievement of the governmental department, it is inevitable that the preparation of rural planning is directly reported to the local township governmental department. The Sanshigang Township rural planning case reflects this point of view. Sanshigang Township Government used power and funds of the district government to promote the rural planning for Dongqu Village and established a mutual commission relationship with the unofficial association and it held a rural planning competition with college student participants. During this process, Dongqu Villagers Committee was only a consultation object and one of the participating members, while the township government dominated the symposium, question answering and expert review processes. Rural planning was not carried out in a bottom-up self-governance manner, neither based on the villagers, nor responsible to the villagers, but a duplication of urban planning model in rural areas.

14.3.3 Neglecting Rural Land Use System in Preparation of Current Rural Planning

In rural areas of China, land is not only the means of production, but also the similar property of village residents, and a key part of their social security. The dual nature of urban and rural land ownership in China determines that the contents and methods of rural and urban planning are different in nature. The technical performance characteristics of urban planning are inclined to be constructions of urban public space, division of different function parts of the city, and the connection of the private and semi-private urban spaces with the public space systems of different hierarchies, scales and forms. In rural areas, land is collectively owned, the right to use belongs to village residents. Their property rights may increase or may be inherited with the change of population. At the same time, the technical performance characteristics of rural planning are subject to the characteristics of land-use right and rural planning cannot start with the public space network system, such as roads, squares and meeting places. The method and content of urban planning in the background of the state-owned land cannot be directly copied to establish feasible solutions to the rural problems. Therefore, many current rural planning do not have practical values and meanings, which are reflected in the following three scales:

(1) For village territorial coverage, land transfer leads to the phenomena of agricultural recession, idle farmland and nearsightedness in industrial development. Under the government-led land transfer, the separation of the land contract rights and operation rights results in the separation of the three powers,[2] which

---

[2] The separation of the three powers mentioned in the context of Chinese land regulation system means the pattern of ownership, contract right and operation right being separated that is developed based on the implementation of collective rural land ownership, stabilization of famer's contract right and invigoration of land operation right.

indirectly leaves most village residents with no land to cultivate and only a small amount of year-end bonus. However, rural planning often does not consider the different interests among individual villagers, villagers committee, land lease operation entity, and township government. The industrial policies proposed, such as agricultural tourism and picking experience economy, seem effective, but cannot solve the mismatch between human and land under the current land regulation system. Take the rural planning competition case of Dongqu Village in Sanshigang Township as the example. After land transfer, the period of operating cycle is 20 years. Due to the consideration for return on investment, operators of Taoxi Fruit Farm and Guaniu Fairyland remarkably reduces the commercial crops, leading to the decline of traditional special agricultural products such as wild peach and watermelon. Moreover, the construction and development in the name of recreation facilities and experience economy restricts the agricultural modernization development (sorted according to the interview recording, No. ZTH-20170731–01). The survey also shows that the land transfer is not based on the voluntariness of all villagers, and large number of villagers feel regret for the misplacement of the transferred land, fading of the agricultural water system and the recession of traditional crops (see Fig. 14.3).

(2) For relations between villages and village territorial coverage, the difference in the ownership of homestead and contracted land after land transfer, as well as the resulting land space separation are often neglected. In addition, it takes for granted that the villages and farmland can be divided as living space and working space according to the urban function zoning. For example, in the case of Dongqu Village in Sanshigang Township, a wire mesh is established between the homestead (residential area) and contracted land (farmland) as a boundary. Villagers can only plant vegetables on the vacant land in front of their cottages. During the interview, we learned that some villagers of Dongqu Village without farming skills can only go to the cities to do some physical works in service industry since they lost their land. It can be seen, that it is meaningless to discuss the promotion of rural industry and cultural renaissance through planning, if rural planning neglects the institutional factors hidden behind the land (see Fig. 14.4).

(3) For internal factors of the village, it disregards the space characteristics of homesteads that are distributed in a scattered manner and blindly copies the contents and methods of urban planning to the villages. We can often see that some rural planning takes the collective land as the state-owned land and scarcely pays attention to the local characteristics of villages, utilizing improper planning and design methods such as squares, residential groups, and road facilities that are specific for state-owned land. In fact, cottages should be constructed by the villagers themselves. Some village collective lands are designed as public space or parking lots as the cities do, which is unreasonable and infeasible. Besides, the so-called compensation mechanism is often ambiguous and impractical due to the unclear land boundaries. In summary, a large amount of rural planning designs do not consider the fundamental system, common sense, and basic conditions of the people (see Fig. 14.5).

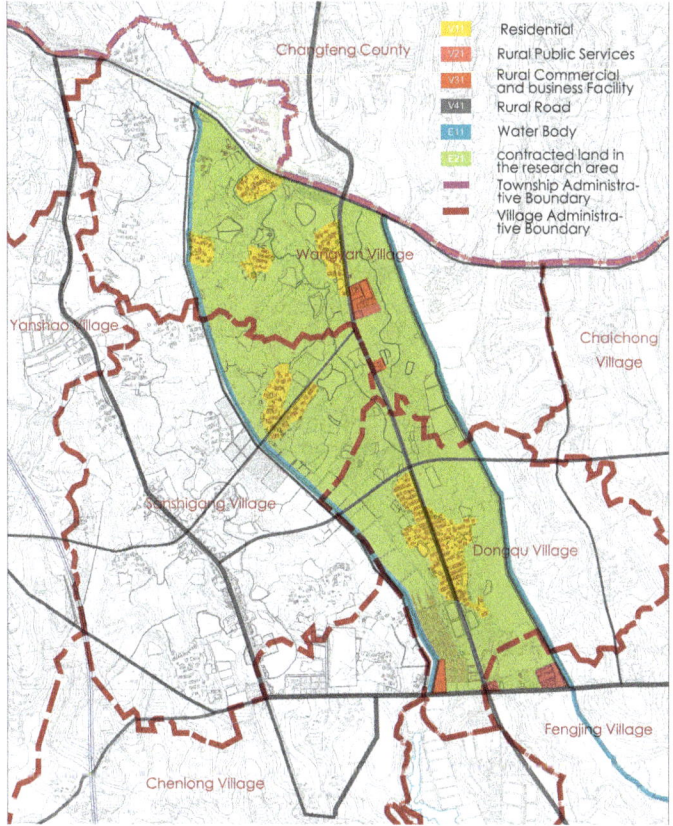

Fig. 14.3 The transaction of land use in township level

14.3.4   Inadequate Attention on Practical Conditions of the Village and Villagers' Perception in Technical Means for Rural Planning and Result Expression

The urban planning has its own systematic expression method and the audiences are fixed, so the policy and implementation information conveyed by the drawings and texts can be widely received and understood. However, the application of these urban planning representation methods to rural planning will obviously encounter serious obstacles and problems. Because the characteristics of rural areas and the target receivers are unique, it becomes a normal state that the rural planning results expressed by the urban planning expression method cannot be understood by the villagers and are deemed nonstandard by the experts. This not only affect the ability to convey planning concepts, but also make it difficult for planning to play a guiding role (Yanfei et al., 2016).

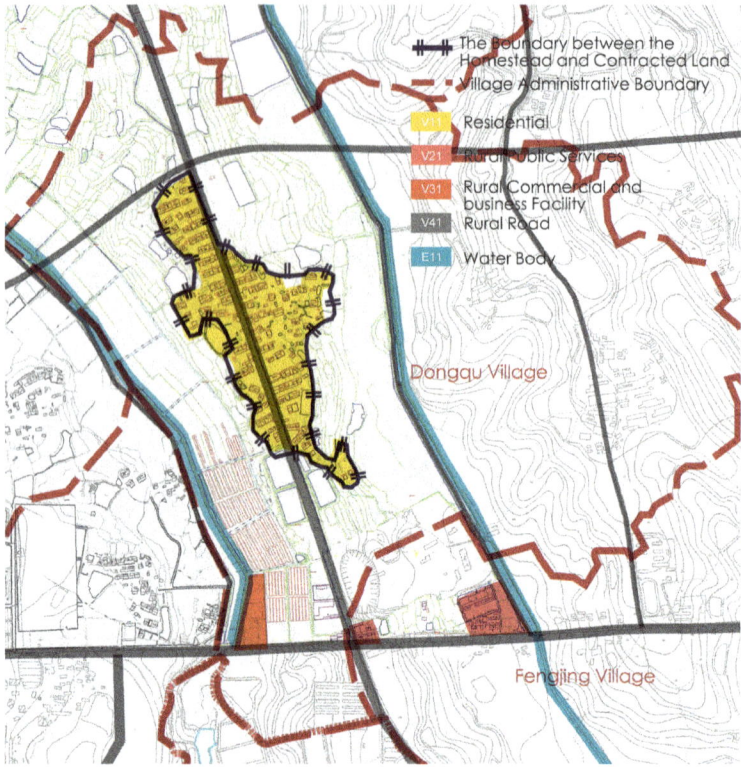

Fig. 14.4 The space segregation between resident area and Farmland

There are three aspects in planning drawings. First, divide the urban areas and rural areas by means of regional planning and according to the general arrangement of urban and rural social economic development and construction; make overall space arrangement of rural residential areas in accordance with the positioning of population and economic industry and by way of movement and mergers of villages (see Figs. 14.6 and 14.7). Second, divide the natural villages within the administrative village into different functional spaces such as production space and living space, and make detailed arrangement on the land use, such as planning structure and functional layout (see Fig. 14.8). Third, use the technical language of detailed planning for residential area construction to propose the space planning for daily lives and public activities, including road system, green land landscape and detailed site design. Meanwhile, some suggestions for developing industries and activating local context are also included (Figs. 14.9 and 14.10).

For communication methods, there are also three problems. First, during the research, the living modes and practical demands of villagers are not well understood because of the interference from the township government. Second, the research is also conducted by way of questionnaires and interviews that are adopted by the

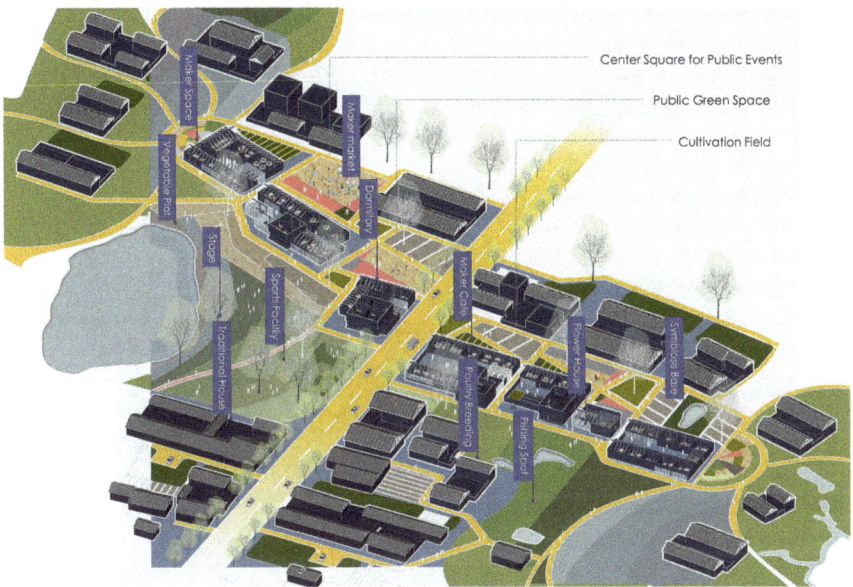

Fig. 14.5 The design of public space in case

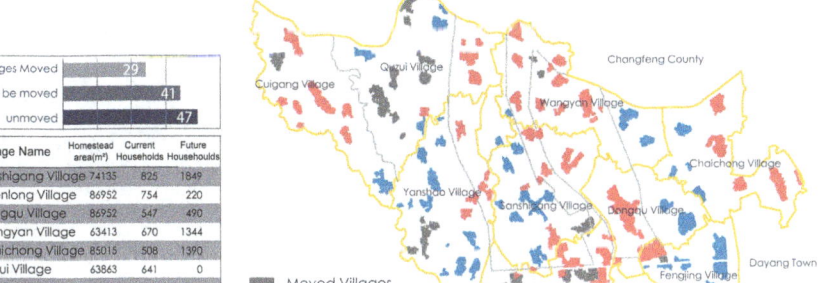

Fig. 14.6 The settlement system planning at township level

urban research, neglecting the difference between villagers and urban residents in understanding methods, comprehension ability and communication skills. Villagers are used to receiving information running from mouth to mouth, and cannot understand the written results (Yanfei et al., 2016) and the questionnaires. Third, research results show that the villagers' wills are not sufficiently respected and documented and some officials and experts are reluctant to consider the problems, opinions, and

Landuse and Spatial Structure Planning

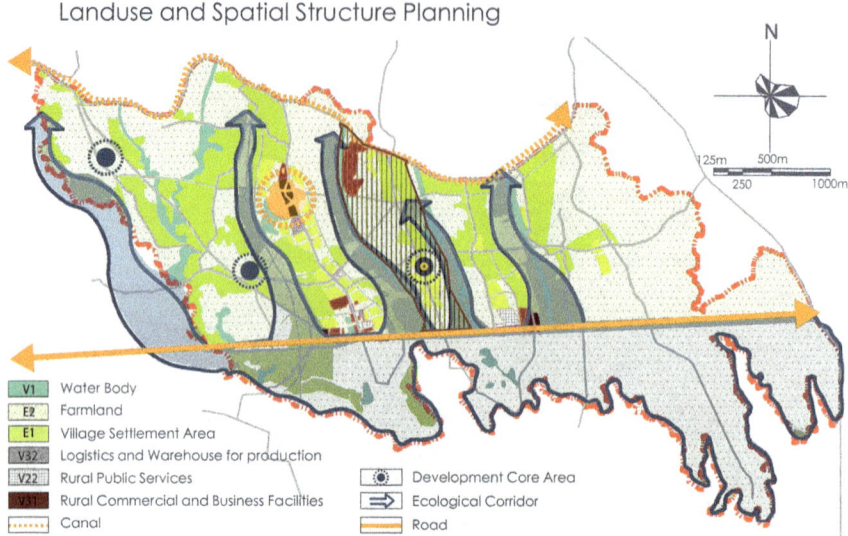

Legend:
- V1 — Water Body
- E2 — Farmland
- E1 — Village Settlement Area
- V32 — Logistics and Warehouse for production
- V22 — Rural Public Services
- V31 — Rural Commercial and Business Facilities
- Canal
- Development Core Area
- Ecological Corridor
- Road

Fig. 14.7 The concept planning at township level

Regional Planning

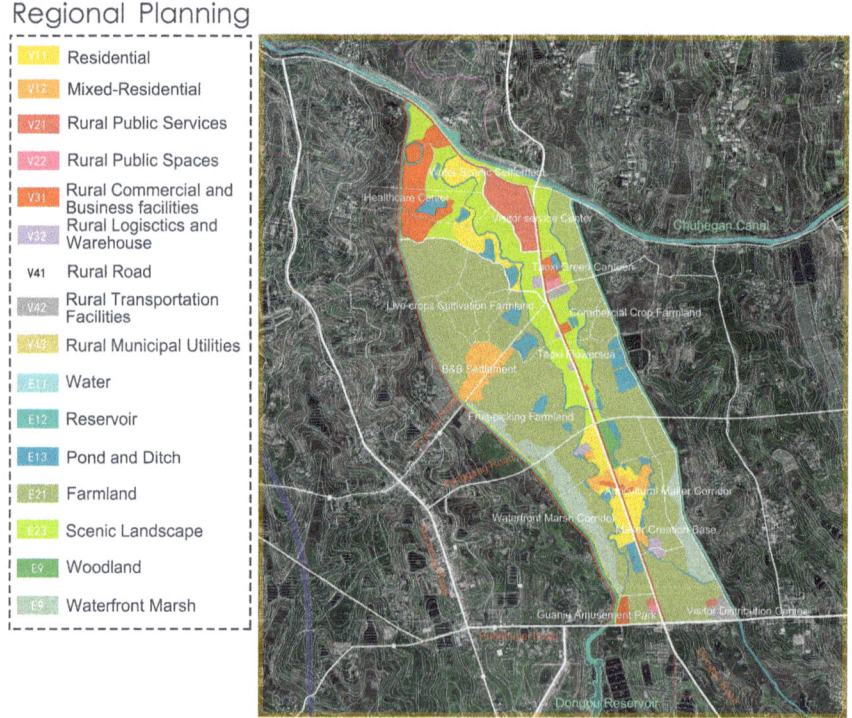

Legend:
- V11 — Residential
- V12 — Mixed-Residential
- V21 — Rural Public Services
- V22 — Rural Public Spaces
- V31 — Rural Commercial and Business facilities
- V32 — Rural Logisctics and Warehouse
- V41 — Rural Road
- V42 — Rural Transportation Facilities
- V43 — Rural Municipal Utilities
- E11 — Water
- E12 — Reservoir
- E13 — Pond and Ditch
- E21 — Farmland
- E23 — Scenic Landscape
- E9 — Woodland
- E9 — Waterfront Marsh

Fig. 14.8 The land use planning at village level

Fig. 14.9 The spatial planning at village level

values reflected by the villagers. In the rural planning competition of Dongqu Village in Sanshigang Township, the achievement documents are competition oriented and only cater to the academic or intellectual thought-processes of urban planning experts and urban management experts, and barely make any effort or exploration to get villagers involved in submitting ideas in the competition.

This kind of rural planning is hardly coherent with actual demands because it only considers the governmental planning management departments and cannot be understood by the villagers; the subjective preparation of construction indicators and contents are inapplicable to the current conditions of villagers due to the insufficient communication with villagers, leading to the waste of government investment (Yaolin et al., 2016). Rural planning deserves more attention to the characteristics of villager self-governance, the differentiated and changing interest demands, complex land ownership, and unclear implementation subject. It may not be favorable for the realization and implementation of urban development strategy to transit gradually from the traditional top-down planning preparation to the attitude of attaching more

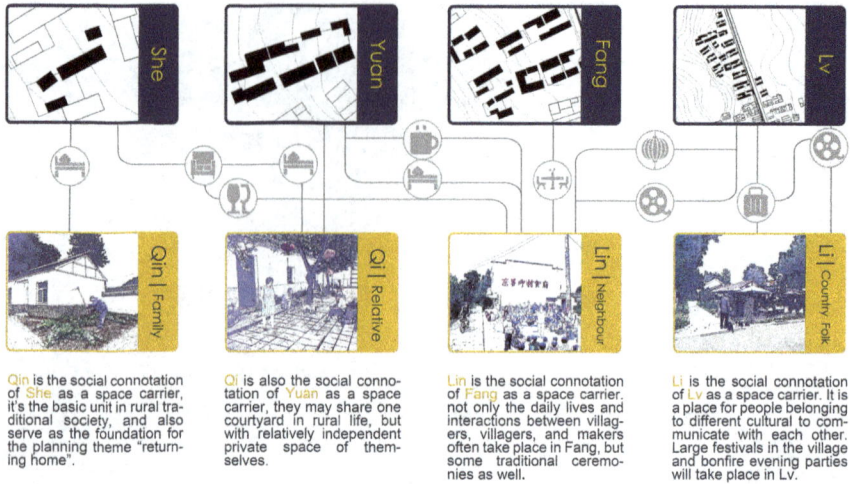

Qin is the social connotation of She as a space carrier, it's the basic unit in rural traditional society, and also serve as the foundation for the planning theme "returning home".

Qi is also the social connotation of Yuan as a space carrier, they may share one courtyard in rural life, but with relatively independent private space of themselves.

Lin is the social connotation of Fang as a space carrier, not only the daily lives and interactions between villagers, villagers, and makers often take place in Fang, but some traditional ceremonies as well.

Li is the social connotation of Lv as a space carrier. It is a place for people belonging to different cultural to communicate with each other. Large festivals in the village and bonfire evening parties will take place in Lv.

Fig. 14.10 The spatial pattern corresponding to rural society

importance on the villagers' standpoint. However, it is favorable for resolving the practical difficulties and preparing for responsible rural planning.

14.4   Problems in Rural Planning Education

Urban–rural planning is never the simple integration of urban planning and rural planning, thus a complete system should be established based on the sufficient understanding of the regional development characteristics of both the urban and rural areas (You & Jianyun, 2017). Therefore, the theoretical knowledge system of rural planning should be expanded based on the theoretical knowledge system of urban planning, and further explorations on the planning design methods are required, since it is far from mature and cannot play an effective social role. The defects of these theories, knowledge and methods should be clearly pointed out to the students, to allow the students to get rid of the low-level imitation and subjective conjecture. The author holds that the following reflections should be made

14.4.1   What Should the Rural Planning Education Teach?

Rural planning should not be taught as the urban residential area planning because rural communities are different from urban residential areas. The above-mentioned difficulties encountered by the contemporary village planning of China are not suffi-ciently taught in the current rural planning education, though they are the essence for rural planning.

First, rural planning is the reconsideration of urban planning. Rural problems are also the urban problems, yet they are two different faces of one coin. Therefore, rural planning education should focus on educating students to fully understand the characteristics of urban and rural areas in the context of China and to establish a new and improved urban planning method.

Second, rural planning education should let the students realize that rural planning is the planning of social groups and settlements different from cities. Rural and urban areas in China do not only show the difference in employment and residence places, but also the basis for dividing identities, which leads to the difference in income, welfare distribution, social security, living mode, living environment and so forth.

Third, rural planning education should help students to understand the particu-larity of the villager self-governance. Villagers self-governance is a crucial form of grass-root democracy and the village collective organization has the right conferred by the Constitution to decide the public affairs and public welfare undertakings within the village. Rural planning education should interpret the guidance rela-tion of the *Constitution, Organic Law of the Villagers Committees of the People's Republic of China, Property Law,* and *Villagers Self-governance Regulations* with the planning and construction rules and regulations such as *Urban and Rural Planning Law.* Meanwhile, rural planning education should clarify the conflicts and contradic-tions between the planning management authorities and grass-root self-governance powers from the perspective of national systems, figure out the real entrusting subject and implementation subject of rural planning, and take them as academic research projects and discussion topics.

Fourth, rural planning education should guide the students to explore and use space language and expression methods that are different from those of urban planning and help students realize that land regulation system affects and decides the space planning language. Rural planning education ought to use targeted and creative space modes and vocabularies applicable to the rural context for different living groups, social relations, land regulation systems and governance mechanisms. At the same time, the communication channels and methods that suit the demands and desires of the villagers should be explored. It is mainly embodied in the important work links such as on-the-spot investigation, will consultation, achievement report and decision forming. In addition, multiple participation modes that are consistent with the characteristics of the villagers should be explored. For example, the aging of population, hollowing out, stay-at-home children, and migrant workers in rural areas lead to complex situations, because some village residents live in the rural areas while others do not. Land property and its social insurance attributes keep residents who are not currently living in the rural areas linked as important "interest-related subjects" who cannot be neglected in the rural planning (Fig. 14.10).

14.4.2 How Should the Rural Planning Education Guide the Students to Establish Value Orientation?

The content of rural planning education shall not only reflect the thinking logic and method system, but also have guidance for developing and establishing the value orientation of rural planning.

First, value orientation decides whether the rural planning serves for villagers committee or basic-level government. This does not question the principal-agent relation between the two parties, but emphasizes the matter of how to select a proper standpoint and form a common action program when the autonomous community and state community have different wills and both wills are inconsistent with the suggestions of the third-party planner group (for example, the government only considers the short-time economic interest, while village residents only want to earn more money).

Second, as the implementation tool of rural revitalization strategy, the decision of rural planning to revitalize agriculture, countryside, or villagers, (or the three at the same time), determines the core connotation of rural planning. Agricultural development can improve the economic situation, rural development can contribute to a livable and enterprise adaptable environment, potentially leading village residents to return to the countryside based on the above changes and expectations, for local civilization can be continued. As the demographic dividend gradually disappears in China, every scholar and official should rationally and effectively coordinate the Rural Revitalization strategy and New-type Urbanization strategy.

Third, serving the countryside or consuming the countryside is also related to the core value orientation of rural planning. How should we conduct rural revitalization, by attracting villager-workers to return to the rural areas to reconstruct rural society, or carrying out industry implantation or industry upgrading according to the consuming demands and taste characteristics of urban residents, just taking the declining rural spaces as a consumption object to bear industry? Should the "gentrification of rural spaces" be appreciated and studied? These questions are worth consideration.

14.5 Discussion: Reflection on the Understanding of Urban–Rural Relationship in China by Rural Planning

The difficulties encountered by rural planning and the bias of planning education are the critical points for us to understand and think about the urban–rural relationship in China.

(1) Land regulation systems and land use planning should be appreciated by urban planning. Rural planning makes the topic of land regulation system widely appreciated in the field of urban planning. It plays an essential role in

understanding the urbanization process in the context of Chinese systems and establishing Chinese rural planning theory.

(2)   Urban and rural governance modes and systems must become the cores of urban planning. Rural planning makes the relation between self-governed territory and state administrative authority-governed territory unprecedentedly important. The think on the issue of whether it is the extension of the administrative power of planning or the demand of self-development of autonomous organizations is the reason for the establishment of rural planning and the implementation of management is conducive to form a correct interpretation towards the core value of urban planning and the professional noumenon.

(3)   The duality of urban–rural societies and spaces should be taken as the object of research and function of urban planning. There are villager-worker problems in urban areas and there are industry problems in rural areas. The two cause and effect each other, together constitute the social economic issue with Chinese characteristics. Rural revitalization may be conducted not only in rural areas, but urban areas as well. For example, by considering how to provide better employment and living environments for migrant workers, and how to guarantee them receive equitable access to social services. New-type Urbanization may be carried out not only in urban areas. For example, it should consider how to promote the equal urban and rural power, fair resource exchange, and balanced distribution of public security. Urban–rural dual structure shows the two aspects of a goal or problem in space and urban planning scholars should pay attention to the realization of urban–rural coexistence, social equity and space justice and the resolution of unbalanced capital accumulation and social demand during the space production process.

The author holds that the intervention of the academic and practical community of urban planning into the rural areas should not be considered as the loss of power by urban planning in urban areas or planner groups working in the rural areas collectively, as some scholars worried about, but should be understood as an exploration process for urban planning to make self-repair and self-improvement of subject knowledge and theoretical framework. The research and substantial resolution of rural problems also urge the urban problems to be identified and resolved. Rural planning is an important chapter and part of urban planning, not a new field or a new subject paradigm separated from the urban planning.

14.6   Conclusions and Prospects

The highly autonomous social system and the "collecting tax according to hectare of land" land regulation system have been developed since ancient times in China. The state has issued a series of laws to specify the self-governance rights of village committee as a grass roots self-governing unit and the collective ownership nature of farmland. Nonetheless, the current rural planning preparation practices and rural planning education shows that the planning realm does not have a systematic under-standing of the particularity of rural problems, leading to the lack of clear and effective paths for rural planning research and implementation. The defects in law sources, entrusting and implementation subject, and technical planning representations and communication modes mean that rural planning can currently, only play a weak role in realizing the national strategy and policy. The planning community should actively create and impart related knowledge and understanding of the urban–rural differences, attach more importance on the institutional factors, think the effect and meaning of rural planning on the contemporary urban–rural relationship in China, and improve the urban planning theory and practice. This must have a long-term impact on the establishment of Chinese urban planning theoretical system.

Acknowledgement Acknowledgements are given to all the members of Tongji University Team participating in the First National Rural Planning Competition for College Students (Hefei Base) held in Dongqu Village of Sanshigang Township, including the students of Zhou Tianyang, Wu Yiying, Han Shuo, Shao Xinyao and Zhang Yujie as well as the instructors of Yang Fan, Luan Feng, and Zhang Shangwu.

References

Binkai, C. H. E. N., & Yifu, L. I. N. (2013). Development strategy, urbanization and income gap between urban and rural areas in China. *Social Sciences in China, 4,* 81–102.

Guosheng, Z. (2009). A study on transformation of peasant workers into urban residents in the perspective of social cost. *China Soft Science,* (4), 56–69.

Jiangang, W. E. N., & Hanzi, W. E. N. (2015). A study on the implementing problem of rural governance and planning. *Modern Urban Research, 4,* 16–26.

Jianrong, Y. U. (2010). Village self-governance: Value and difficulties-and on the revision of "Law of the People's Republic of China villagers committee." *Study & Exploration, 4,* 73–76.

Ke, Z. H. O. U., & Jing, G. U. (2017). Planning and development control of traditional villages under villagers' self-governance—the case of post-disaster reconstruction of Caojia Village. *Urban Planning Forum, 2,* 87–95.

Kun, H. (2011). Influence on new-generation migrant workers citizenization from rural land regu-lation system and institutional innovations. *Research of Agricultural Modernization, 32*(2), 196–199.

Linxiu, Z., Qiang, L., Renfu, L., Chengfang, L., & Sigao, L. (2005). China's rural public goods investment and its regional distribution. *Chinese Rural Economy,* (11), 18–25.

Qiang, Y. A. N. (2012). Township planning management system study from constitutional viewpoint. *Planners, 10,* 13–17.

Qing, X. U., Shichao, T. I. A. N., Zhigang, X. U., & Ting, S. H. A. O. (2008). Rural land regulation system, land fragmentation and village resident's income inequality. *Economics Research, 2,* 82–92.

Sihong, W. U. (2003). The stability of village elite's benefit game and power structure. *Journal of the Party School of the Central Committee of the CPC, 7*(1), 39–43.

Weiyi, L. I. (2004). On the game between township government and village self-governance. *Social Sciences in Yunnan, 7,* 11–13.

Xi, C (2011) *Three basic problems of rural development in China.* Beijing:People's Daily press.

Yanfei, L. I., Wenjun, L. I. U., & Hao, Y. A. N. G. (2016). Exploring the expression of practical village planning results in the new era - based on the needs of different participants. *Jiangsu Urban Planning, 1,* 25–29.

Yaolin, M. E. I., Shanshan, X. U., & Hao, Y. A. N. G. (2016). Pragmatic rural planning compilation practice. *The Planner, 1,* 119–125.

Yiqiang, L. I. U., & Jun, H. U. (2014). The village household system tradition and its evolution: A rediscovery of the basic institutional forms of China's rural governance. *Study & Exploration, 1,* 53–59.

Yong, G., & Xiyun, K. (2005). Governmental function and policy decision of city plan. *Shanxi Architecture,* (2), 132–133.

You, Z. H. O. U., & Jianyun, Z. H. O. U. (2017). Farmer identity system, market economy, and rural planning. *City Planning Review, 2,* 94–101.

Zhongyuan, C., Mei, H., Degang, D. (2016). Inheritance and practice of rural planning. *China Architectural Education,* (2), 67–72.

Chapter 15
Obstacles and Opportunities for Characteristic Town Development in Central China Area: Hubei Province Case Study

Shuting Yan, Toshikazu Ishida, Mamiko Fujiyama, and Xilin Zhou

15.1 Background

By the end of 2015, China's urbanization level reached 56.1%, and urbanization entered the middle and late developmental stages. Zhejiang province was the first Chinese area to implement characteristic town policy (Huang and Tan, 2017). The development strategy a town chooses is an important aspect to realizing productive transformation in Zhejiang province in terms of resources, ecology, and labor (Zeng and Cl, 2016). Currently, Tongxiang of Wuzhen ("Internet town") and Yunqi town of the West Lake district have become the model towns of Zhejiang province. Their successful experiences have received worldwide attention, and have played leading roles for the development of small towns nationwide.

On June 4th, 2015, Zhejiang province officially released the first list of provincial characteristic towns which comprised of 37 towns in 10 districts of the province. The second list of 42 towns was announced in early 2016 (Yangming, 2016). On December 24th, 2015, President Xi Jinping announced for a national study of Zhejiang to be conducted to build a characteristic town. This action made characteristic town in Zhejiang a hot topic throughout the country.

In May 2016, the NDRC (National Development and Reform Commission) stated a goal for China to construct and support nearly 1,000 characteristic towns by 2020. These towns would feature vibrant and distinctive industries like leisure tourism, trade logistics, modern manufacturing, education, and science and technology. Additionally, these towns would embrace traditional cultures and cultivate beautiful livable towns which ultimately would improve the level of construction and the quality of development (Zhang, 2017). During the same year on October 14th, the NDRC announced the list of the first completed 127 characteristic towns. In August 2017,

S. Yan (✉) · T. Ishida · M. Fujiyama · X. Zhou
Wuhan Institute of Technology, 693 Xiongchu Street, Wuhan, China
e-mail: Nancyan803@outlook.com

School of Civil Engineering and Architecture, Wuhan Institute of Technology, Wuhan, China

© Springer Nature Switzerland AG 2021
W. Li et al. (eds.), *Human-Centered Urban Planning and Design in China: Volume I*,
GeoJournal Library 129, https://doi.org/10.1007/978-3-030-83856-0_15

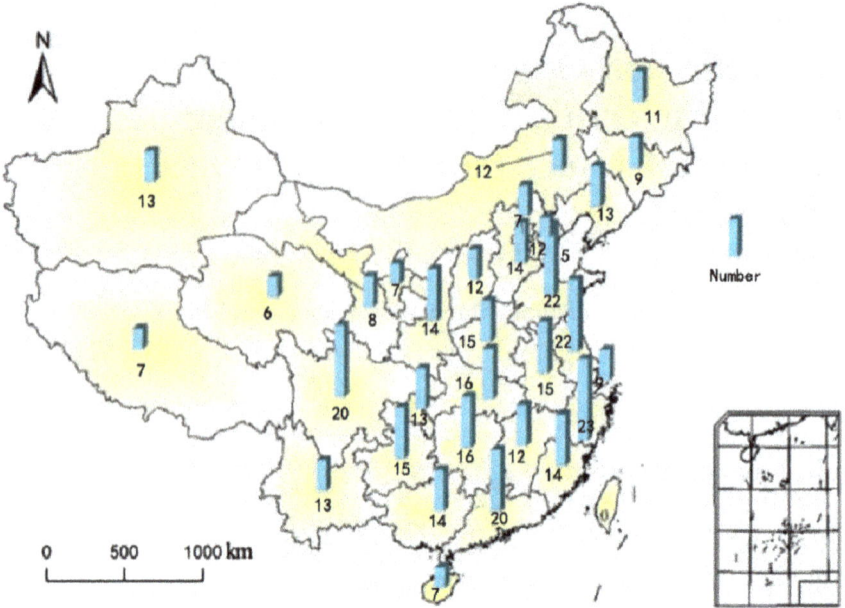

Fig. 15.1 The map of the number distribution of characteristic towns (first batch + second batch) (*Data Resource* Ministry of Housing and Urban–Rural Development of the People's Republic of China)

the second group of 276 characteristic towns was announced (Fig. 15.1). In view of the cumulative number of two batches of the 403 announced characteristic towns, Zhejiang province had the most (23). Other provinces that had 20 or more characteristic towns were Jiangsu, Shandong, Sichuan and Guangdong. The 403 characteristics towns were predominately in the more economically developed eastern provinces and the southwest areas of China. Although the southwest area's economy is underdeveloped, it is rich in tourism resources.

While the east and southwest areas are thriving with characteristic town development, the large economic gaps in the central and western regions of China cause them to be far behind in the quantity and quality of characteristic town development (Yonghua, 2017). Not only should the central and western regions model the successes of the coastal regions for characteristic town development, but also learn from other developed countries' strategies.

First, this paper analyzes the policy background of the construction of the characteristic town and outlines the basic construction of the characteristic towns in the whole country and Hubei provinces. Next, we review issues Hubei Province faces in terms of industrial structure, development and construction, policy support, ideological consciousness, and infrastructure. Finally, the development of Hubei's characteristic towns is discussed to provide a reference for the construction of characteristic town in other less-developed places.

15.2   Research Objectives

After comparing the number of characteristic towns, provincial GDP rankings, and the provincial GDP ratios of the primary, secondary and tertiary industries, we found there is limited research on characteristic towns in Hubei province. Given that Hubei province is representative of underdeveloped central and western regions in China, our research objectives can be summarized in the following aspects:

(1)   The comparison of the relationship between the number of characteristic towns and provincial GDP ranking.

Hubei province, which is economically better than the most provinces in China, had the 7th best GDP ranking in 2016 among 31 total provinces. The relationship between number of characteristic towns and provincial GDP ranking can be expressed through a Univariate Nonlinear Regression generated by the function of Regression Analysis in Excel where X is Provincial GDP ranking; Y is the number of characteristic towns in every province; R^2 is 0.68, which means the correlation is high.

$$Y = 0.0023X^3 + 0.1229X - 2.2619X + 25.947$$

This result means that the more GDP a province generates, then the more characteristic towns it will have. As shown in the Fig. 15.2, the trend can be divided

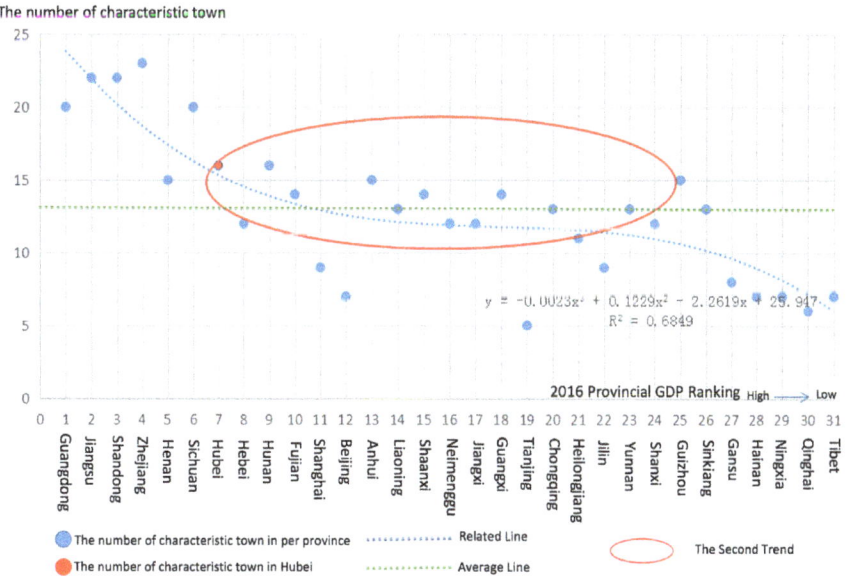

Fig. 15.2   The relationship of the number of characteristic towns and provincial GDP ranking (*Data Resource* National Bureau of Statistics of China. Retrieved September 15, 2017)

into three parts: (1) The number of characteristic towns a province among the top 6 provinces in terms of GDP, it drops considerably but begins to smooth out. (2) The second grouping, which contains the provinces with the 7th to the 22nd highest GDPs, have relatively the same number of characteristic towns. (Fig. 15.2). Hubei province which has the 7th highest GDP, has a total of 16 characteristic towns, which is the same as Hunan (9th highest GDP). This observation highlights that Hubei has a lot of potential.

(2)   The comparison of the relationship between provincial GDP ratios of the primary, secondary and tertiary industries.

Through comparing the GDP of the primary industry, the secondary industry, and the tertiary industry to the total GDP of the 31 provinces, we found that the main difference is between the secondary and tertiary industries (Fig. 15.3). The fluctuation between these two levels of industry has two stages. The output value of the secondary and tertiary industry of the top five provinces (Guangdong, Jiangsu, Shandong, Zhejiang, and Henan) are much higher than that of Hubei. With this finding, we suggest that the number of characteristic towns is determined by the output value of the secondary and tertiary industries. Based on the analysis of the developed secondary and tertiary industries in Jiangsu and Zhejiang area, Hubei should consider how to position its secondary and tertiary industries for success, so that they can form their own advantage of competitiveness similar with that of the top five provinces.

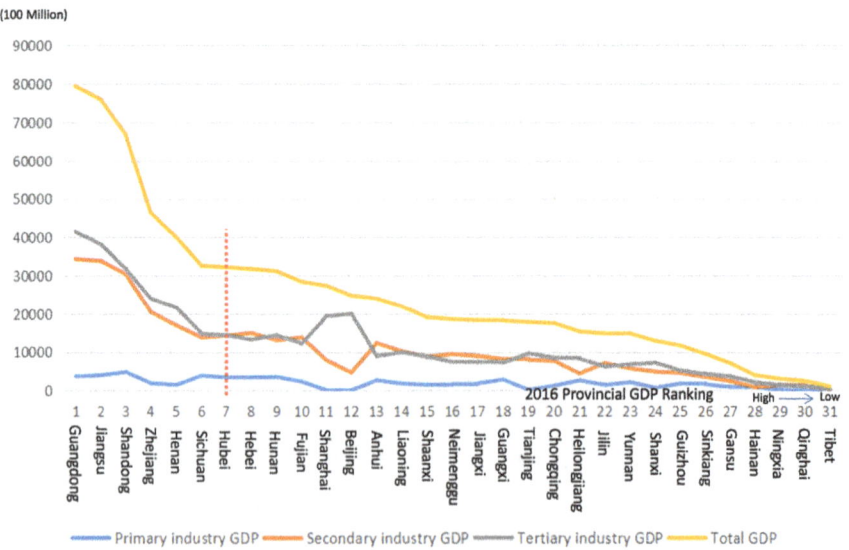

Fig. 15.3 The relationship among provincial GDP ratio of the first, secondary and tertiary industry (*Data Resource* National Bureau of Statistics of China. Retrieved September 15, 2017)

(3)   The existing research of characteristic towns in Hubei is insufficient.

Research about characteristic towns is mainly concentrated in the developed areas of Zhejiang. Any research conducted on characteristic towns in Hubei focuses on how it can learn from successes in Zhejiang province. However, we suggest Hubei should be individually studied to best understand its unique problems so we can find feasible, localized solutions to these problems.

15.3   Analysis of Industrial Characteristics and Development Dilemma of Hubei Characteristic Towns

In October 2016, Hubei province had five characteristic towns in the first round of the announced 127 towns. These five towns are Dragon Spring town, Wudian town, Flight Enthusiast town, Chujing Culture town, and Zen town. These towns have unique name based on their developmental goals.

For example, the town of Longquan in Yichang City is focused on developing as a famous liquor and eco-friendly town. This town has successfully implemented the use of green hills, clear water, and created a harmonious and livable environment in villages. Wudian Town in Zaoyang City adheres to the "Garden City" and "Integration of Construction and Production Cities" concepts. These areas focus on industrial features, living environments, traditional culture. They aim to create a town that is industrially strong, has beautiful architecture, is inviting for tourists and has charming qualities. Jinghe Town is guided by the "General Aviation New City, Green Ecology" aspect. The new district will build national research and development and manufacturing bases for special-purpose aircrafts and general aviation aircraft, aviation emergency rescue centers, on-board air test-testing centers, and position itself as a well-known aviation exhibition and experiential destination.

In August 2017, the NDRC ministry published the second list of characteristic towns of the 276 towns, Hubei province had 11 towns on the list (Table 15.1). Through the comparison of the 16 characteristic towns in Hubei province and based on the most important element of urbanization development, we identified that Hubei's main obstacles for the development of characteristic towns is the lack of innovation industries, the relatively low GDP level, the insufficient support for continual industrial development, and population loss.

15.3.1   Lack of Innovation Industries

In the 16 towns, Hubei province has formed six major types of town characteristics (Fig. 15.4): (1) Leisure tourism (43.75%); (2) Modern agriculture (18.75%); (3)

Table 1 The list of first batch and second batch of characteristic towns in Hubei province

Num.	Area	Characteristic Town Name	Characteristic Town Type	Time	Longitude	Latitude
1	Longquan town,Yichang city	Dragon spring town	Industrial Development	2016	110°29'E	30°43'N
2	Wudian town,Xiangyang city	Wudian town	Historic Culture	2016	112°46'E	31°58'N
3	Zhanghe town,Jingmen city	Flight enthusiasts town	Leisure Tourism	2016	112°04'E	30°57'N
4	Qiliping town,Huanggang city	ChuJing culture town	Historic Culture	2016	114°39'E	31°27'N
5	Changgang town,Suizhou city	Zen town	Leisure Tourism	2016	112°58'E	31°33'N
6	Weishui town,Jingzhou city	Sports resort town	Leisure Tourism	2017	111°37'E	29°59'N
7	Zhaojun town,Yichang city	Zhaojun town	Leisure Tourism	2017	110°45'E	31°14'N
8	Xiongkou town,Qianjiang city	Lobster town	Modern Agriculture	2017	112°47'E	31°18'N
9	Pengchang town,Xiantao city	Countryside town	Modern Agriculture	2017	113°30'E	30°16'N
10	Xianrendu town,Xiangyang city	Green circle town	Environmental Protection	2017	111°43'E	32°15'N
11	Huiwan town,Shiyan city	Gong tea town	Traditional Industries	2017	109°49'E	32°08'N
12	Guanqiao town,Xianning city	New material town	Industrial Development	2017	113°57'E	29°55'N
13	Hongping town,Shenlongjia	Eco-tourism town	Leisure Tourism	2017	110°25'E	31°40'N
14	Yuxian town,Wuhan city	Gardening town	Leisure Tourism	2017	113°59'E	30°31'N
15	Yuekou town,Tianmen city	Eco-farming town	Modern Agriculture	2017	113°05'E	30°30'N
16	Moudao town,Enshi city	Tourist town	Leisure Tourism	2017	108°41'E	30°25'N

Data Resource Ministry of Housing and Urban–Rural Development of the People's Republic of China. Retrieved April 17, 2018.

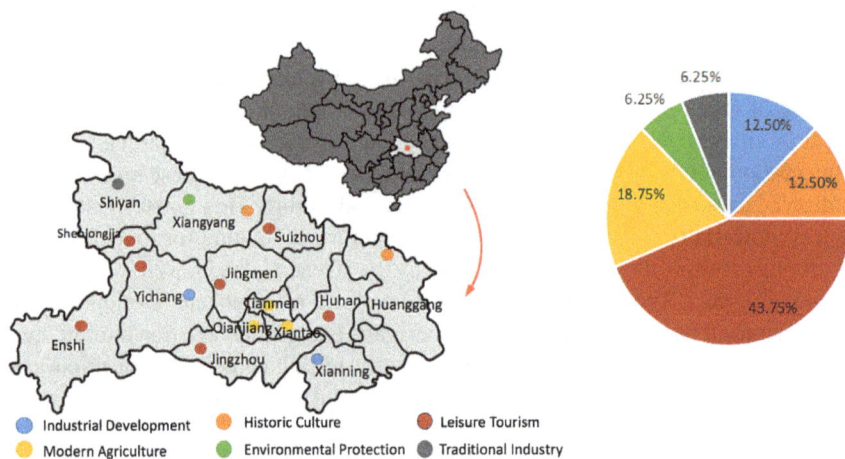

Fig. 15.4 The Characteristic town type distribution in Hubei province

Historic culture (12.5%); (4) Industrial development (12.5%); (5) Environmental protection (6.25%); (6) Traditional industry (6.25%).

The top characteristic towns in Hubei province are tourism and modern agriculture, and they are not seeing many characteristic towns focused on the innovation industry. The reasons for a lack of innovation industries are due to unbalanced regional development, the limited political, economic and cultural resources, and lacking competitive industry. However, other larger cities or provincial capitals have innovative industries influencing younger people to locate in these areas, whereas smaller towns cannot attract these innovative, young people as successfully. Due

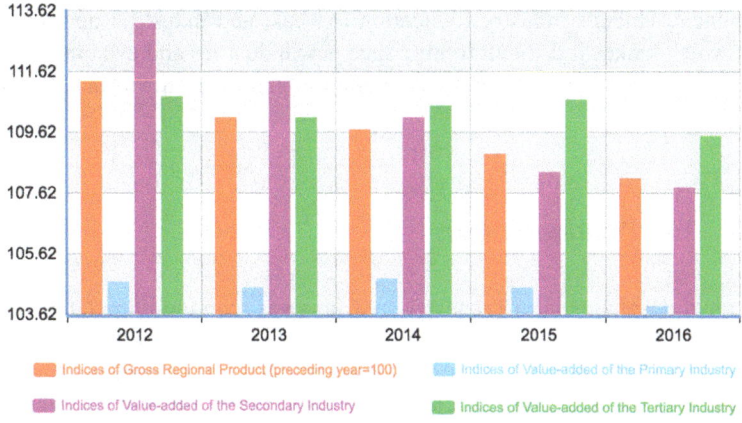

Chart 15.1 Analysis of industry growth index in Hubei province (2012–2016) (*Data Resource* National Bureau of Statistics of China. Retrieved September 15, 2017)

to this predicament, the smaller cities, which have lost the support of the youth innovation community, are likely to fall into innovative talent desertification.

15.3.2 The GDP is Low and Industrial Development Momentum is Insufficient

China has entered a stage of high quality development. Although the assessment of GDP tends to be diluted, this indicator is still crucial. According to preliminarily data from the National Bureau of Statistics in China, Hubei province's GDP ranked seventh nationally. This equates total of 3.229791 trillion yuan in 2016 meaning it became one of nine provinces with more than 3 trillion yuan in China. However, a big gap still exists between the GDPs' provinces in Guangdong, Jiangsu and Shandong which are ranked in the top three (Fig. 15.3).

Japan's demographic data from 1960s which began to distinguish between urban and rural populations based on the municipal and county population. The data of urban and rural populations in all counties (grouped by five year increments) can be traced back to 1898, and are highly comparable in both longitudinal and lateral directions providing a good statistical analysis sample for empirical research.

According to the analysis of industry growth index from 2012 to 2016 in Hubei province, the GDP growth index is decreasing year over year, and the primary industry and the tertiary industry growth index is fluctuating in the most recent five years. Additionally, the growth rate is rather stable, and the second industry growth index is decreasing which means the relative advantage of Hubei province has fallen by 4.1% in the past five years. This data reveals that industrial development is insufficient in Hubei characteristic towns, and the phenomenon of "semi-urbanization" is prominent. Hubei's industry foundation is weak, and industrial development is slowly which makes it difficult to introduce new industries and even more difficult to retain industries. These issues are common for other small towns in China as well (Chart 15.1).

15.3.3 The Population Loss

In the rapid development of urbanization, China has seen the importance of population and technology exceed the significance of investments for the first time. All cities and provinces in China have shifted their focus from increasing investments into increasing population and talents which is reflective of a higher-level competition model. Some developed cities and provinces have taken the lead on achieving population and talent growth; however, most cities and provinces remain ignorant or incompetent to this shift and they are facing population loss.

Hubei province is one of the provinces with active population migration and mobility. The top five provinces with the largest population losses are: Anhui (8.9 million), Sichuan (7.77 million), Hunan (6.5 million), Jiangxi (5.18 million), and Hubei (4.87 million). The total net population loss of these five years out of the provinces combined is more than 4.5 million, accounting for about 10% of the population of the provinces.

In Hubei province, the number of inter-provincial outflow dropped from 5.89 million in 2010 to 4.97 million in 2016, while the number of provincial inflow population rose from 1.01 million in 2010 to 1.49 million in 2016. Although the inflow is growing faster than the outflow, the inflow is much smaller than the outflow. (Fig. 15.5).

Fig. 15.5 Analysis of population flow in various cities of Hubei province

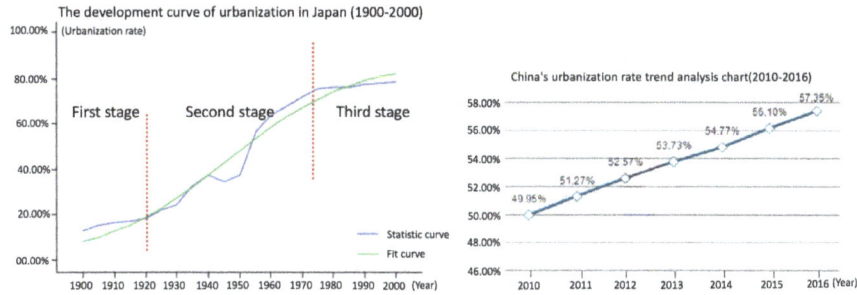

Chart 15.2 Analysis the relationship of urbanization rate between China and Japan

15.4 Case in Study from Northeast of Japan

15.4.1 The Comparison Between Japan and China's Urbanization Stage

Dr. Jianjun WANG in the study on the unitary characteristics and regional differences of fast urbanization stage in Japan shows that Japan entered the second state of urbanization in the 1920s. Within 50 years, the rapid development of urbanization was completed, and the third stage of urbanization development began in the 1970s. During the rapid development stage of urbanization (the second stage) in Japan, the average annual rate of urbanization increased by nearly one percentage point, a very fast increase (Wang and Wu, 2007) (Chart 15.2).

The rapid development of China's urbanization started later than Japan, but the speed of development was fast. In the mid- and late-1970s when the urbanization level was around 17%, China entered its second stage of urbanization development with predictions of entering the third stage in the early-2030s when urbanization level would be about 63% (Jianjun, 2015).

To determine urban and rural populations, Japan uses municipal and county population demographic data. The data of urban and rural populations in all counties (grouped by five-year periods) can be traced back to 1898. The data are highly comparable in both longitudinal and lateral directions which provides a good statistical analysis sample for empirical research. Compared to Japan, China is a large country with different natural conditions and resources, economic bases, folk cultures, and habits. These features and the urban development scene experience more regional differences. According to China's urbanization rate trend from 2010 to 2016, development is occurring at a rapid pace that is equivalent to Japan's second stage of urbanization development. Therefore, China is still a developing country, and the experiences and lessons of Japan's urbanization process will serve as a good reference point for China's future urbanization development.

15.4.2 The Similarity of Difficulties in Urbanization Process Between Northeast Japan and Hubei Province in China

In the case of regions like Hokkaido, Kanto, Central, Kinki, Tohoku, Shikoku, and Kyushu in Japan and through our urbanization analysis using several key factors like economic development, population loss, and traffic development, we found that Japan's urbanization development process in the Tohoku region has the highest reference value for Hubei province.

(1)   Economic development status.

Japan's northeast metropolitan circle (Tohoku region) including Aomori ken, Iwate ken, Miyagi ken, Akita ken, Yamagata ken, Fukushima ken and Niigata ken, has a total cumulative population of all these areas of about 13.5 million, a GDP of about 40 trillion yen, a land area of about 73.718 square kilometers, and contains four cities, like Sendai city, that has more than 300,000 people. Conversely, Hubei province has a total area of 185.9 square kilometers, is about 2.5 times that of northeast area of Japan, and a population of about 59.02 million people. In 2017, the GDP was 65,295.5 billion yuan. According to 2011 GDP data of different regions in Japan, there is a significant gap among the economic development of Chubu region, the Capital region, and Kinki region (Chart 15.3). This disparity is similar with the economic development status of Hubei province in China.

(2)   The relationship between population loss and urbanization.

The population trend has been negative from 1955 to 2010 in the seven prefectures in the Tohoku region (Chart 15.4).

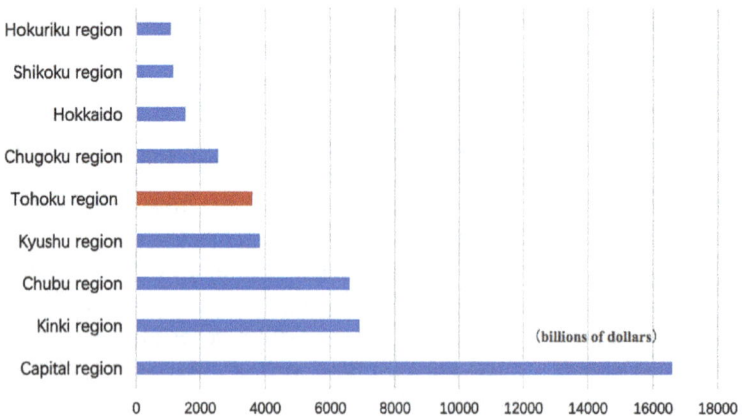

Chart 15.3 Comparative analysis of GDP in different regions of Japan (2011) (*Data Resource* from Northeast region summary in Japan(data), National Land Planning Bureau)

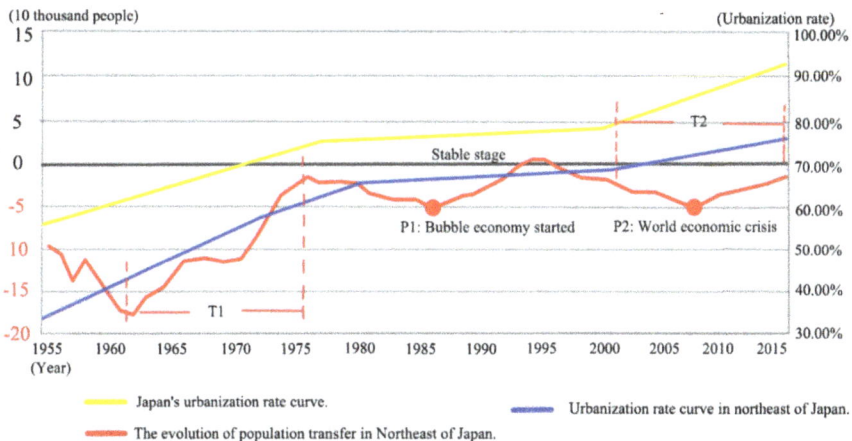

Chart 15.4 The relationship between population loss and urbanization (*Data Resource* from World Bank Open Data: https://data.worldbank.org/)

After 1975, the provinces in northeast Japan completed the second phase of urbanization. The urbanization rate was at about 52–68%, which is similar with that of Hubei's currently status. In this stage, there was a problem of population loss which is also like the current situation of urbanization in Hubei province.

With the urbanization rate increasing in Hubei, the big cities will continue to grow, and the smaller cities will gradually shrink. The larger cities will begin to absorb the surrounding smaller cities in the process of their expansions. Some less developed areas, such as Tohoku region in Japan and Hubei area in China, will see their populations leaving for bigger cities facing greater urbanization. This process is the population loss urbanization is causing smaller cities. Population loss effected the Tohoku region in Japan the most from 1960 to 1975. China is now seeing the peak of this phenomenon compared to any time in its history since it is in a rapid urbanization period. The development of characteristic towns helps slow this population loss trend.

The comparison of data in Chart 15.4 shows that the urbanization rate was negatively correlated with population loss in northeast region of Japan. During a period of steady urbanization, population loss of the northeast region was volatile due to the impacts of an economic crisis from Japan becoming a financial economy. It was initially thought that the urbanization period was over, and there was little to no development in northeast Japan and a large loss of population from people moving to Tokyo.

After Japan signed the "plaza accord" in 1985, the Japanese yen appreciated substantially, and real estate began to gradually develop. The housing price in Tokyo's metropolitan circle rose, and the population in the northeast region of Japan began to recover. Therefore, the population loss of P1 is caused by the financial crisis, which has no relationship with the urbanization rate. The economy recovered after the P1 point and urbanization continued to increase., and the population of the northeast region began to migrate to the metropolis. Our study concludes that urbanization rate growth is negatively correlated with population loss in the northeast of Japan without interference from other factors. The higher the urbanization rate, the lower the loss in population. If small towns want to develop, they must have population growth, otherwise the urbanization rate will be very slow during an economic crisis.

The periods of P1 and P2 indicate that the growth of urbanization was small compared to growth during Japan's economic crisis in the 1990s and the global economic crisis in 2008. Instead, the two economic crises reduced the population of Tohoku region. The two growth inflection points in northeast Japan were driven by the government's revitalization policies: Tohoku development promotion plan (First) in 1958 and Tohoku development promotion plan (Fifth) in 1999.The government's guiding policy was the decisive factor for urbanization growth in northeast Japan. The most successful policies were the first and the fifth Tohoku development promotion plans. Since the fifth revitalization plan, the northeast region has experienced an increase in population loss, so the first revitalization plan appears to be more successful. Perhaps due to the Japan's aging population and the lack of innovative employment to attract talent, the population in the northeast was more attracted to Tokyo.

15.5 Discussion

By comparing the statistical data of the urbanization process in Japanese prefectures to that of development indicators in Chinese provinces, we find a relatively high similarity in the urbanization process between the northeast of Japan (1960–1975) and Hubei province (2010–2030).

Japan completed its rapid development stage of urbanization several decades ahead of China. The above analysis shows that the problems faced by northeast Japan in their rapid growth stage of urbanization are similar with that of Hubei province. Since then, effective policy measures have been adopted, and China can learn from the experiences (Table 15.2).

From the policies and measures used in northeast Japan, there are several key takeaways for successful construction of Hubei characteristic towns (Chart 15.5):

(1)   Realize the county-county railroad, expand the traffic to the towns.

In order to improve the popularity of characteristic towns in Hubei province, we must first strengthen the construction of rail transit and improve the convenience and accessibility of transportation. Construction of Shinkansen in northeast Japan began in 1982 which was relatively late compared to other parts of Japan. The rapid and convenient rail transit system such as JR and subway offers Japanese residents more convenient options for travel, which is not as much of the case in China's big cities. Additionally, Japan's transit network allows for better linkage between towns, and routes geared toward tourists can be leveraged to attract more tourism. For example,

Table 15.2 The policies and measures for the progress of urbanization in northeast of Japan

Year	Policies and measures	Year	Policies and measures
1957	Tohoku Development Three Laws	1982	**Tohoku Shinkansen opened**
1958	Tohoku development promotion plan(First)	1986	Tohoku Development Co., Ltd. privatization
1962	Comparison of nationwide comprehensive development plan	1987	Comparison of nationwide comprehensive development plan (Forth)
1964	Tohoku development promotion plan(Second)	1989	Tohoku development promotion plan (Forth)
1969	New comparison of nationwide comprehensive development plan	1992	Yamagata Shinkansen opened
1971	Mutsu Ogawara development plan	1997	**Akita Shinkansen opened**
1972	Northeast Vertical Expressway open	1999	Tohoku development promotion plan (Fifth)
1977	Comparison of nationwide comprehensive development plan (Third)	2008	Second National Land Formation Plan (National Plan)
1979	Tohoku development promotion plan (Third)	2010	Hachinohe ⇒ Shin Aomori Tohoku Shinkansen opened

Data Resource from Northeast region summary in Japan(data), National Land Planning Bureau

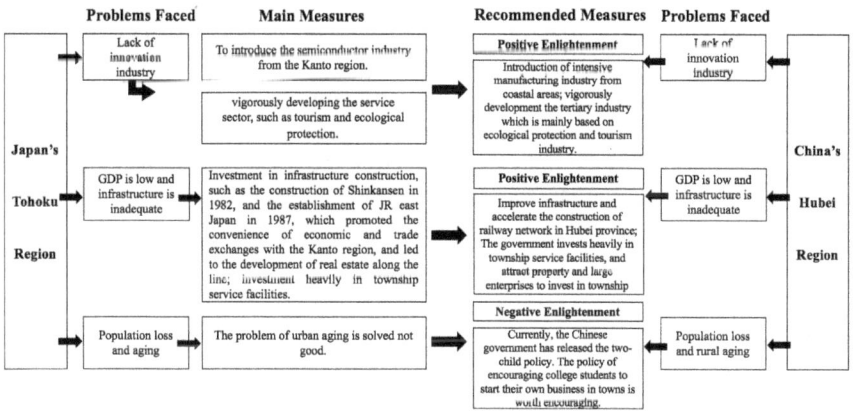

Chart 15.5 The enlightenment of urbanization in northeast of Japan to China's Hubei region

in the cases of Matsushima and Zao which are famous scenic spots in Tohoku region, it is very convenient to get there by JR using this mode of transportation.

Although rail transportation is now developing relatively well (e.g., China High-speed Railway, K/Z/T line Railway, Intercity Railway and Metro) and the Hubei region has begun to build Wuxi high-speed railway and Wuhan 8 + 1 city circle, the construction of intercity railway also needs to be further improved and extended to realize county-county railway networks.

(2)   Encourage college students to start careers in towns and villages.

It is necessary to control population outflow to the larger cities in Hubei province. For example, in the case of northeast region of Japan, many people, especially highly-competent people, graduated from college in the northeast and then moved to Tokyo to start their careers. Hubei should encourage university students from cities like Wuhan, Xiangyang, and Yichang to stay in these towns by developing competitive employment opportunities in these county and towns. The retention of industry and high-quality talents are keys to the development of successful characteristic towns.

(3)   Hubei should strive to develop second, tertiary industries with a focus on the tourism industry.

According to the Pie-Clark theorem, with the improvement of economic development and per capita income, a proportion of primary, secondary industries gradually transforms to the tertiary industries. Additionally, the proportion of labor also shifts from the primary industry to secondary industry to the third industry, which follows the universal law of employment structure change. In the development of characteristic towns, Hubei province is given priority to eco-tourism and environmental economy, and the second industry is complementary. Long-term, the proportion of future economic output from the secondary industries will decrease, and their development potential will shrink. The development of intensive industries such as technology-intensive industries, energy-intensive industries can also negatively effect the environment. This sentiment to develop a tourism industry to attract foreign travelers is echoed by the current policy in northeast Japan. In April 2018, President Xi Jinping put forward the policy of protecting the Yangtze river economy zone, developing tourism and protecting the environment, and encouraged eco-economy and environmental tourism.

15.6   Conclusion

Characteristic-town policy is helping big cities' concentrate industry distribution to restrict the excessive expansion of urban suburbanization and accelerate the counter-urbanization development process. This study uses 16 national characteristic towns in Hubei province that represent the central and western regions of China to analyze obstacles of characteristic town development. These obstacles include that lack of innovation industries, population loss, relatively low GDP, and insufficient industrial development. By comparing the statistical data of Japanese prefectures' urbanization processes and development indicators to that of Chinese provinces, we found that the two stages of rapid urbanization in northeast Japan (1960–1975 and 2000-present) represent the next stages of urbanization in Hubei province.

The two growth periods in the northeast region of Japan were driven by the government's revitalization policy: Tohoku development promotion plan (First) in 1958 and Tohoku development promotion plan (Fifth) in 1999. By analyzing the gains and losses in the first and second stages of the rapid growth of urbanization in the northeast region of Japan and by examining the first and fifth policies of the government's revitalization, we found some good suggestions for Hubei province. These suggestions include: (1) In terms of traffic network, Hubei region needs to achieve the transportation benefits the county-county railroad and expand the traffic to the towns. The construction of inter-city railway also needs to be further improved and extended; (2) In terms of attracting talent, Hubei officials should encourage college students to continue working in its towns and villages post-graduation. The retention of industries and high-quality talent are the key to the development of characteristic towns; (3) In terms of industrial structure, Hubei should strive to develop second, tertiary industries focused mainly on tourism.

Acknowledgements This study was supported by the Chinese Scholarship Council (Grant Number 20160950012).

References

Huang, Y., Tan, J. (2017) An overview of special town and its construction principles and methods. *Journal of Guangxi Cadres College of Economic and Management*, (29), 93–98.

Jianjun, W. A. N. G. (2015). The unitary characteristics and regional differences of fast urbanization stage in Japan. *International City Planning, 30*(03), 59–65.

Ministry of Housing and Urban-Rural Development of the People's Republic of China. Retrieved 17 April 2018. http://www.mohurd.gov.cn/wjfb/201708/t20170828_233078.html

National Bureau of Statistics of China. (2017) Retrieved 15 September 2017. http://data.stats.gov.cn/easyquery.htm?cn=E0103

Northeast region summary in Japan(data), National Land Planning Bureau. http://www.mlit.go.jp/singikai/kokudosin/sankou2.pdf

The construction of small towns in Hubei province focuses on cultivating. (2013) singles champion. *Northern Architecture, 3*(01), 86.

Tohoku region: Ministry of Land, Infrastructure and Transport, Tohoku Regional Development Bureau, Outline of the Tohoku region area planning. http://www.thr.mlit.go.jp/kokudo/kuikig aiyo/kuikigaiyo.html

Wang, J., & Wu, Z. (2007). 1950 years after the world's major countries urbanization: path analysis with the class group. *Journal of Urban Planning, 172* (6), 47–53.

World Bank Open Data. (2017). Retrieved 6 December 2017. https://data.worldbank.org/.

Yanming, S. H. A. N., Huijia, M. A., & Wenjie, S. O. N. G. (2016). Cultivating cognition and interpretation of small towns across the country. *Small Town Construction, 11*, 20–24.

Yonghua, H. A. O. (2017). Analysis of regional differences in characteristic towns and the path of creating characteristic towns in less developed areas. *Enterprise Economy, 36*(10), 171–177.

Zeng, J., & Cl, F. (2016). Construction of featured towns under the background of new urbanization. *Macroeconomic Management*, (12), 51–56.

Zhang, H. (2017). Theory and practice innovation on the construction of featured towns. *China Famous City*, (01), 4–10.

Chapter 16
Research on the Shaping of Landscape in Rural Cultural Heritage Based Areas Using the Optimal Solution Model

Yu Guo, Zhenya Chen, Lingqing Zhang, Jing Yan, Wenfeng Fu, Ying Cao, and Xiaohong Tang

16.1 Introduction

Currently, a wave of rural construction and reconstruction has arisen throughout China with the strategy for the revitalization of villages. However, there is often the conflict between project elements and design intent in the process of rural construction (Hu et al., 2017). The results directly caused by this problem are: the lack of cultural heritage, the lack of characteristics of rural construction style, and the difficulty of having a sense of belonging and environmental identity. However, reasonable solutions are subject to the constraints of cost and other factors that make it difficult to achieve.

The study point for the Dujiangyan essential irrigation area is relatively diverse. Wentao Yan et al. expounded on the value and the ecological wisdom of human settlement of the Dujiangyan essential irrigation area from the perspective of the historical comparison method (Yan et al., 2017). Min Zhang, Feng Han, et al. analyzed the composition and development context of the Dujiangyan cultural landscape from the perspective of the historical landscape and cultural landscape of the city (Zhang, Han & Xu, 2016). Huimin Wang, et al. evaluated and summarized the evaluation system of sustainable development in the Dujiangyan essential irrigation area (Wang, 1999). For the protection of rural cultural landscapes, Lihui Peng made a summary of Guangxi's rural cultural landscape classification and evaluation system (Peng, 2017). Donghua Guo also explored the inheritance and protection of traditional rural cultural landscapes in real cases (Guo, 2013). There has also been in-depth elaboration on

Y. Guo · Z. Chen · L. Zhang · W. Fu · Y. Cao · X. Tang (✉)
School of Architecture and Urban-Rural Planning, Sichuan Agricultural University, No.288 Jianshe Road, Dujiangyan, Chengdu 611830, Sichuan, China
e-mail: 41344@sicau.edu.cn

J. Yan
School of Business, Sichuan Agricultural University, No.288 Jianshe Road, Dujiangyan, Chengdu 611830, Sichuan, China

© Springer Nature Switzerland AG 2021
W. Li et al. (eds.), *Human-Centered Urban Planning and Design in China: Volume I*,
GeoJournal Library 129, https://doi.org/10.1007/978-3-030-83856-0_16

the connotation of the multi-value system and rural landscape and the dual nature of rural cultural landscapes (Jin, 2011; Li, 2011). Dongning Zhang also classified and summarized the heritage of traditional village cultural landscapes (Zhang, 2017). Zhijun Fang, Yuzhi Fan, Fei Xue made a detailed analysis and description of the settlement patterns, environmental characteristics, current conditions, problems, and cultural essence of the Linpan in western Sichuan (Fang, 2012; Fan, 2009; Xue and Zhu, 2013). The optimal solution model is good at analyzing complex problems and widely used in life. It also convinced enough in accurate and quantitative analysis.

These applications all have a common mathematical model: one or more variables can be controlled and are usually subject to some practical restrictions. By controlling these variables, some other variable can reach the optimal result (Meerschaert, 2015). However, the coupling of perceptual scoring and rational analysis is difficult, and the optimal solution model has not been applied to design landscape of rural cultural heritage. Using the optimal solution conceptual model to quantitatively analyze and get the minimum cost that satisfies the constraint condition will make the result directly and objectively, and this also provides the theoretical support and methodology for shaping the landscape of rural cultural heritage which balances the lower cost and higher aesthetic value, better protection of cultural landscape heritage simultaneously.

16.2 Materials and Methodology

16.2.1 Background Information of Study Area

Shiyang Town was selected as the study area which is located in the south of Dujiangyan City, 16 km away from Dujiangyan city center and covers an area of 49.6 square kilometers. Located at N36.55°E103.33°, the town is high in the north and low in the east, with a height drop of four thousandths of a meter. Shiyang Town is located in the lower reaches of Dujiangyan City, 635 m above sea level. It is a subtropical humid climate in the Sichuan Basin. The climate is mild, with an average annual temperature of 17.1 °C, abundant rainfall, long sunshine, long frost-free periods, and four distinct seasons. The perennial dominant wind direction is from the northwest. The industrial structure is dominated by agriculture, with a total cultivated area of 5.86 million mu. The main industries are high-quality grain and oil cultivation, flower seedlings, and ligusticum striatum (Fig. 16.1).

The town boasts abundant water resources, including 4 natural rivers, more than 10 branches, and multiple canal drainage channels, as well as a number of agricultural drainage channels. The Shagou River, Heishi River, and the Zouma River pass through the town and the Jinma River is located at the eastern border of Shiyang Town, providing favorable conditions for agricultural irrigation. Shi-yang Town is also rich in rural culture with personalities such as Madame Huarui, and also the farming culture like ligusticum striatum superior origin, Orchid superior origin and

Fig. 16.1 The location of Shiyang town and Essence Irrigation Area in Dujiangyan. *Image source* Dujiangyan Administrative Division Map colored by the author

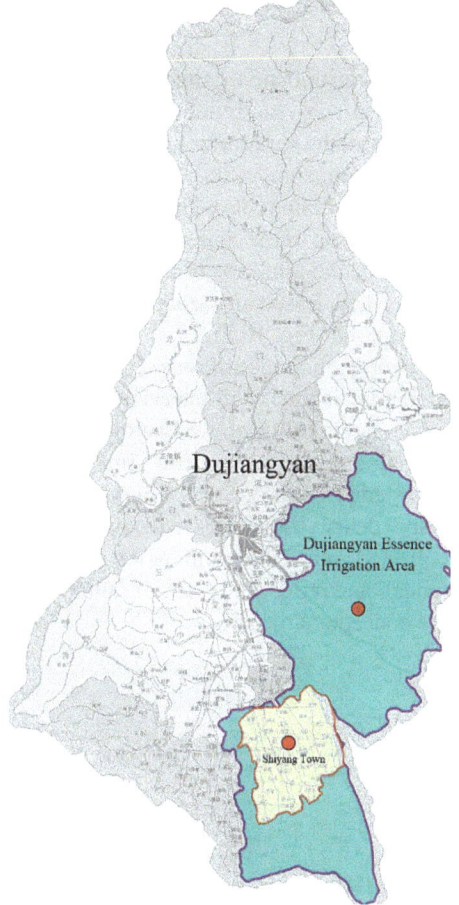

the rice and fish are interconnected. It also has Chen's wood carvings and the Mazu Temple. The irrigation area generally refers to an area with reliable sources of water and channels for the introduction, transmission, and distribution of water. It is a product of human economic activity and it is a semi-artificial ecosystem composed of reservoirs, channels, fields, and crops. In the Dujiangyan Essence Irrigation Area, a large number of water networks consisting of branch channels, canal channels, agricultural channels, and furrows have been constructed to meet the growing population and to maintain a good living environment and a stable society (Willmott, 1989). There has been a large number of rural settlements along those canals and culverts.

16.2.2 Model Construction and Verification

This paper will first collect basic data through field surveys, collate and conduct statistical analysis, refine the shaping elements of landscapes in the rural cultural heritage of the study area, use the analytic hierarchy process to determine the weights, and determine the value of the assessment when the rural cultural landscape heritage of the unit changes through questionnaire surveys. The optimal solution is determined by the impact of the changes in the rural cultural heritage of the unit on the evaluation value, the the total cost is calculated and the weight values of the different measures are verified. Based on the argumentation process, the optimal solution concept model is proposed (Fig. 16.2).

In order to reduce the error, the author used the expert scoring method for analytic hierarchy analysis and interviewed the local residents with the questionnaires. This combination of the above two methods avoids the subjective influence of the expert scoring and the respondents' limitations on the local culture and the professional knowledge of the critics. The subjective influence of the expert scoring and the respondents' limitations will lead to the deviation between expected results and collected data. This combination is not only more credible, increases the rigor of the model, and reduces the risk of failure but also improves the robustness and rigor of the conceptual model.

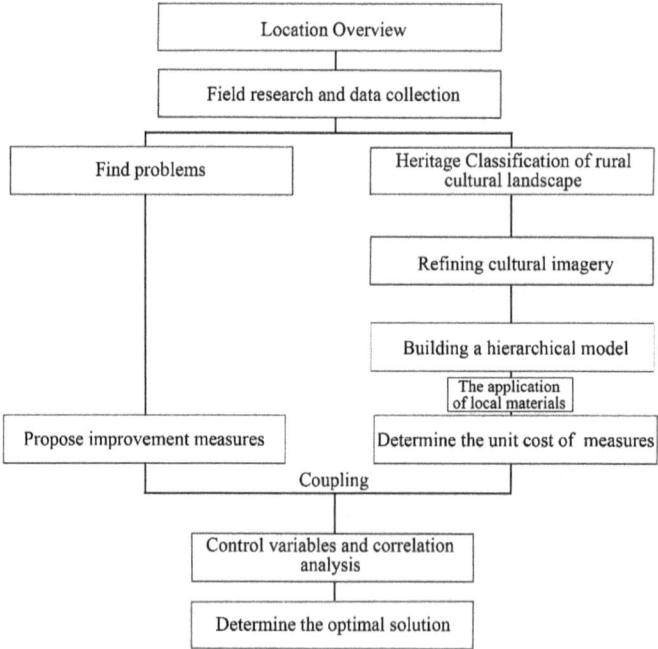

Fig. 16.2 The model of optimal solution. *Image source* the author

The model construction is based on the following preconditions: there is no major change in local policies during the landscape shaping process, the local market price has not changed significantly, the same material has the same price for the same area, materials are collected and calculated on-site and shaping measures are based on widely used local measures.

16.3 Results and Analysis

16.3.1 The Development of Rural Cultural Landscape Heritage

Rural Cultural Landscape Type

"Linpan" is a kind of settlement method for rural residents in West Sichuan. It is usually based on surnames (clan) as a colonial unit. The form is a scattered distribution pattern and is a typical natural village. There are only a few households in small Linpan and hundreds of households in large Linpan. Linpan is generally composed of forest gardens, house buildings and the surrounding cultivated land. The entire house is hidden among tall trees such as phoebe Zhennan, cypress, and low bamboo forests. Most of the Linpan are surrounded by ditches, forming an idyllic picture of a fertile environment, dense forests, Foot Bridge and running water (Lu, 2012). Therefore, Linpan was selected as the starting point to investigate the existing situation. According to the research of site photos of the existing rural cultural landscape, this paper extracts the elements of landscape composition for typical cultural landscapes and divides them into six imageries: characteristics of traditional residential features, characteristics of canal features, the shaping of under-canopy space, the characteristic of open space landscapes, religious architecture style and typical rural agricultural landscapes (Fig. 16.3).

The main problems of the six rural cultural landscape imageries are that the paving materials are generally very simple and lack of diversity and rural cultural characteristics. The local villagers generally do not pay enough attention to them. This often becomes a neglected element of the landscape. There are often gaps in cultural awareness in the local villagers' inheritance process. Most of them only regard it as a symbol of old things and are not aware of historical protection. In the newly renovated residential area, the courtyard aspect ratio is very close to the traditional column and tie construction architectures. In the newly renovated dwellings, the gables are often painted with column and tie construction elements, and there are also some roofs and ridges continuously on the original buildings. The overall environment within Linpan is still dirty and messy, the building style is not uniform, and the construction of many small paths and other infrastructure is not perfect. Many roads are muddy, smell badly and the overall style needs to be improved. The ornamental and practical value of bamboo has been taken into account in Linpan. It acts as a visual center

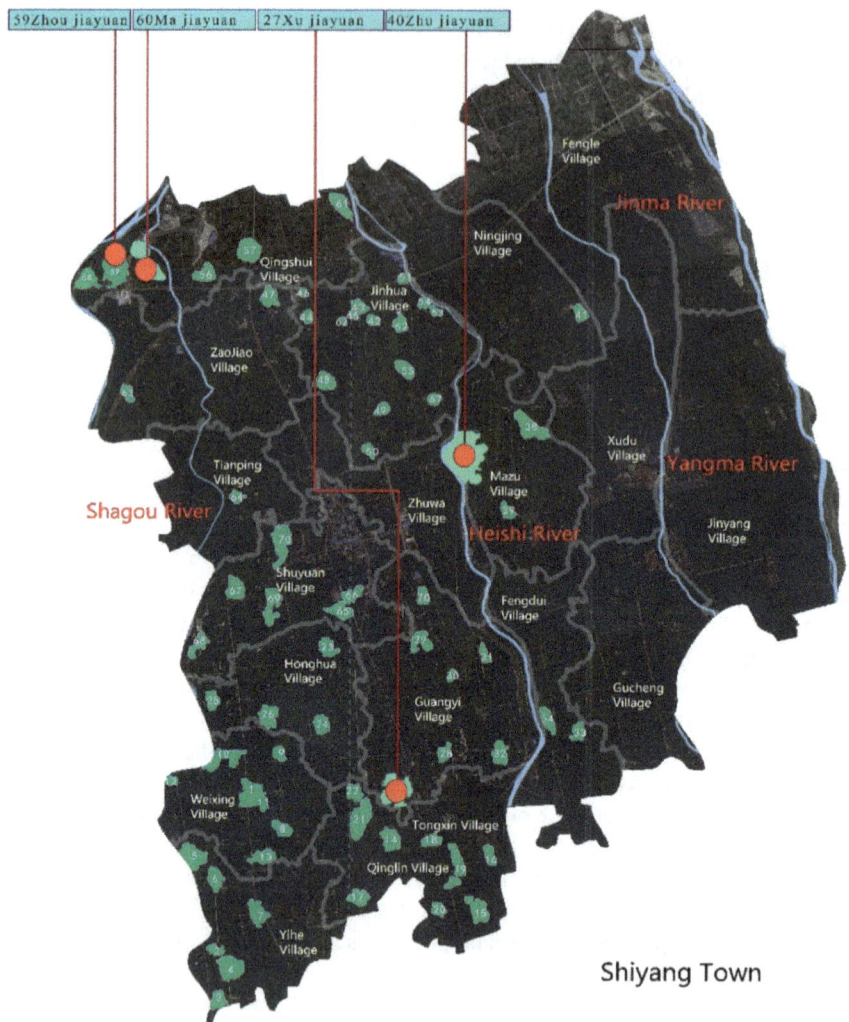

Fig. 16.3 Linpan distribution, river position, research point. *Image source* Google Map Satellite Images which the author colored

and framework for the composition of open space and under-canopy space. The field of vision is severed by dense trunks, creating chaotic scenes without an attractive sight. The building materials of the canal system are made of pebbles that can be seen everywhere on the shores of the local rivers. They are easily accessible and neatly arranged to form a unique order and beauty. The shortcomings from the lack of a signage system are particularly prominent. Different communities are often not identified, and people who come here often lead to disorientation. The locals' familiarity with the environment has led to the obvious neglect of the signage. In the assessment of landscape features above, there are many features and cultural images such as folk art specialties which have not been inherited and known, and they have not been rationally and effectively integrated into the shaping of rural cultural landscapes. There is also a serious lack of publicity for the town and the village's culture. People who are new to this place are likely to be uneducated. The religious architectures need protection and renovation, and the environment needs to be cleared up and repaired. Although there are temple fairs and other events, the education of religious architecture and Mazu Template culture still needs to be improved. Through the measurement of more than 50 photographs, the D/H average of street open space is 1:7.33, and the horizon is narrow and long, forming a unique aesthetic sense in combination with the forest trees. However, this unique scene has yet to be refined and strengthened.

The extraction of rural cultural landscape heritage precedents needs to meet three constraints: the location in the countryside, the definition of a cultural landscape, possessing a history of and currently being a rural cultural landscape retained from generation to generation. The key images extracted from the rural cultural landscape image of the study area through the analysis of the existing conditions which satisfy the above three constraints are: bamboo imagery, imagery of folk houses in western Sichuan, pebble imagery, Chuanxiong imagery, Mazu culture, and Madame Huarui. These images form a rural cultural heritage landscape together with regional features through different combinations.

The Characteristics of Rural Cultural Landscape Evolution

(1) **Dynamic and cyclical**

The methodology of cultural landscape emphasizes that culture is the motivation and landscape is the result (Sauer, 1925). The rural cultural landscape heritage is centered on the village and based on the agricultural economy. It is a type of cultural landscape heritage formed by the harmonious symbiosis between man and nature. It is a dynamically developing landscape heritage type and a cultural space that fully reflects the characteristics of rural culture. The rural cultural landscape not only includes the natural environmental factors of the village but also the factors about the humanities visible in the village. It also includes the gradual transformation of local people. In the course of the natural environment, the important cultural elements such as the environmental outlook, moral outlook, and social values are all integrated in order to form a complete and rich rural cultural landscape (Zhang, 2013). In the process of rural development, rural cultural landscapes are constantly changing in the dynamic development era due to changes in production and life styles, and the integration of new cultures. Because culture is the driving force, landscape is the result, and rural landscape has created a new culture. The new culture has created a new rural landscape, so the process of this change is cyclical again (Fig. 16.4).

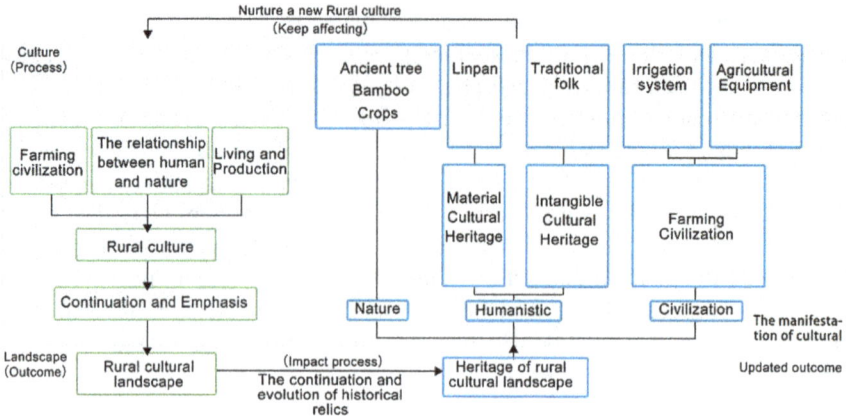

Fig. 16.4 Model of the development of rural cultural landscape heritage. *Image source* the author

(2) **Heterogeneity**

The heterogeneity of the rural landscape reflects the rural biodiversity and a stable landscape ecosystem. In rural landscapes, heterogeneous landscapes refer to landscape complexes consisting of pastoral mosaics, Linpan, woodlands, water bodies, and town neighborhoods that are dominated by grain, oil, and vegetable nurseries. On the small scale, Linpan has various types of areas such as dryland, paddy fields, woodland, residential areas, which crisscross well-developed corridor networks such as roads, ditches, and rivers, forming a complex landscape ecosystem. In terms of agricultural landscapes, China's intensive farming tradition of intercropping, crop rotation, and three-dimensional agriculture, has resulted in a landscape with a small area, a wide variety of crops, and a variety of highly heterogeneous landscapes (Chen, 2012).

(3) **The Unity of Functional and Cultural Relics**

The Dujiangyan Essence Irrigation Area provided the environmental basis for people living here to carry out production and life. People need to grow crops here and they need to build canal systems. Because of the existence of rivers, riverside pebbles have become easily accessible native materials. The use of cobblestones to build a canal system can save raw materials, and the texture of the cobblestone itself adds a rough and rustic aesthetic. The earth piled up in the excavation channel can provide materials for laying roadbeds, so the canal system is often on both sides of the road. Bamboo has many advantages such as a fast growth speed, high practical value, beautiful appearance, etc. The fast growth speed means that the firewood needed for life can be obtained more quickly. Bamboo's stature, high structural strength, and corrosion resistance can be achieved which make it a widely used building material. The lush bamboo creates under-canopy space to provide the visual center and open space for unique aesthetic value. Nurseries and crops provide wide field of vision and

Table 16.1 The ordering weights of elements in the scheme layer for landscape shaping measures

Options	Weights
Column and tie construction (C3)	0.3460
Bamboo planting (C1)	0.2608
Traditional folk mural (C8)	0.1452
Traditional architecture reconstruction (C4)	0.1059
Bamboo chair (C2)	0.0522
Chuanxiong planting (C7)	0.0425
Pebble paving (C6)	0.0279
Ancient tree maintenance (C5)	0.0195

(Evaluation Consistency Ratio: 0.0902; Weights for "Assessment of Cultural Heritage of Rural Heritage in Shiyang Town": 1.0000; λ max: 6.5681.)

farmland landscape while meeting production needs. The unique structural beauty of the residential architectures in western Sichuan and the convenient access to raw materials showing a unique local characteristics, reflecting the traditional farming civilization here, has become a unique cultural heritage.

The rural cultural heritage landscape is a practical starting point. Due to its practicality, it has survived with a long history and it has a pristine beauty in constant evolution. Therefore, the shaping of rural cultural landscape heritage should also focus on practicality combining with beauty.

16.3.2 Analysis of the Elements of Cultural Landscapes in Rural Areas

Rural Cultural Landscape Imagery Weights

After the extraction of the six imageries, AHP is introduced into the decision-making level, the middle level and the program level. The determination of program level considered the feasibility of implementation, the value of cultural images, and the use of local materials. The software is used to build a hierarchical model and expert scoring with scale types (Tables 16.1, 16.2, 16.3, 16.4, 16.5, 16.6, 16.7). A list of matrices and calculated weights is found in Table 16.1 and 16.2.

Rural Landscape Layer Analysis

The correspondence between the decision layer, the middle layer and the plan layer is drawn as a chart, and the weight of each layer is marked on the model chart, as shown in the following Fig. 16.5.

Table 16.2 The ranking weight of elements in the 1st middle tier on the cultural relics

Middle layer elements	Weights
Western Sichuan folk image (B2)	0.4037
Bamboo Image (B1)	0.3129
Mazu Culture (B5)	0.1061
Madame Huarui (B6)	0.1008
Chuanxiong Image (B4)	0.0531
Pebble Image (B3)	0.0234

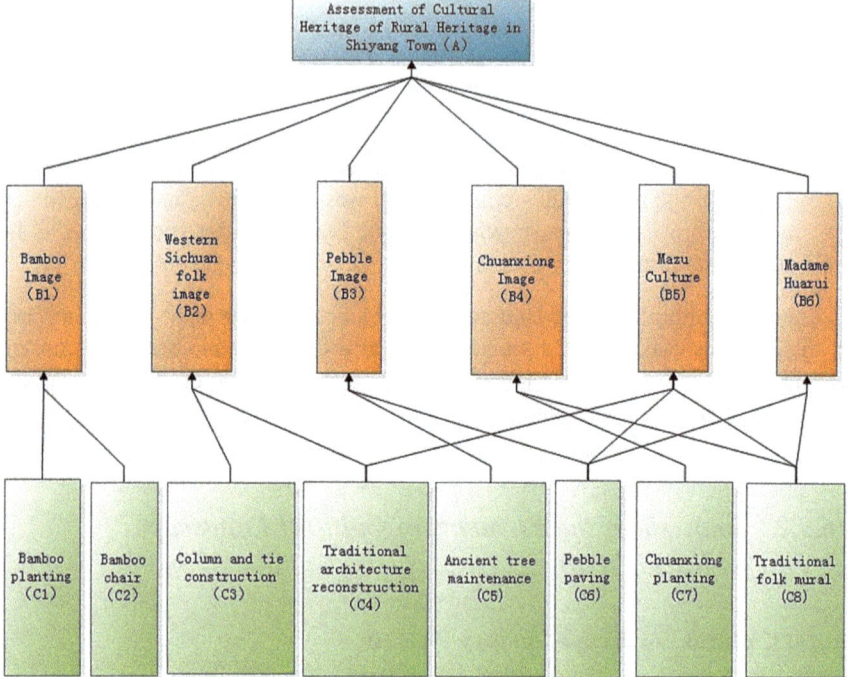

Fig. 16.5 Analytic hierarchy process model. *Image source* the author

The criteria for qualified experts are as follows: long-term contact with local rural landscape planning and design work, participation in actual projects, lived in the local area for a long time, had a rich research experience, and had an understanding of the local situation. Through the expert scoring method for the matrix calculation, the construction of the hierarchical model and the analysis of the calculation of the weight values, the use of the bucket image and the cultivation of the bamboo forest are the two most important weights. The weights of the building restoration and the poster are relatively close. This provides a certain reference value for the actual shaping of rural cultural landscape heritage. However, expert scoring does not represent the completion of the optimal solution model, but also requires further verification.

16.3.3 Optimal Solution Model Construction and Demonstration

Selection of Model Indicators

Traditional dwelling houses that are preserved in good condition were selected as the sample. Zhou Jiayuan of the 12 group of Qingshui Village was selected to conduct research and analysis, and eight improvement measures were incorporated into the actual transformation measures to determine each one. The unit cost of the measure, using the control variable method, sorts the increase and decrease of the measure in the photo, and categorizes the different pictures. Through the analysis of the current environment and a reasonable design method, the simulated photo with multiple solutions is created, and try to improve the authenticity of the picture. Through the rectification and transformation of the courtyard space at a microscopic perspective, the purpose of shaping landscape is achieved. The Zhou family's courtyard shows in the data the forestry plate that can be contiguously considered. The Zhou family's courtyard has a traditional dwelling house. The courtyard space is calculated to cover an area of 400m². There are 65 ginkgo and nanmu trees with a DBH of more than 10 cm. The hand-made specialty is woven bamboo and has the typical rural area characteristics of cultural landscape heritage.

Through field surveys and comprehensive calculations, the unit cost of the eight measures is listed in Table 16.3.

Among them, the bamboo from the planting of seedlings added labor costs and maintenance costs after the average of 20,563 yuan per hectare, an average of 2192 bamboo per hectare, an average of 146 acres, and the cost of 1391 yuan. In the photos, the bamboo planting method is based on a unit cost of acre, which averages 3.5 per acre. Therefore, the cost of the bamboo from planting to growing is about 397 yuan. While the ancient and famous trees are often planted under the trunk, the volume of the pebble mixed concrete mortar is calculated.

Table 16.3 Eight kinds of measures of element unit costs

Measures	C1	C2	C3	C4	C5	C6	C7	C8
Unit Cost	397 [block]	76 [each]	850 [m]	650 [m²]	582 [m³]	80 [m²]	875 [acres]	800 [m²]

Model Index System Construction

Three typical photos from traditional courtyards are selected to represent the interior of the courtyard and the exterior of the courtyard. Two of them are inside the courtyard and one outside the courtyard. Then, the control variable method was used along with different measures to shape different photos (Table 16.4).

Estimating Landscape Costs Based on the Optimal Solution Model

Based on the above random sampling survey in Shiyang Town, a total of 139 valid questionnaires were completed. The questionnaires used a correlation analysis method and a combination of graphic and textual comparison scores. The score scaled 1 to 9, 9 representing the most beautiful, 1 representing the least beautiful. Then, a large number of questionnaires are issued to ensure the number and intuitive nature of the survey. Finally, SPSS software statistics was utilized to get the results in Tables 16.5, 16.6 and 16.7.

According to the trend analysis of the average value of the scoring, the scores of a1–a6, b1–b4, and c1–c5 gradually increased from low to high. Overall, the environmental transformation was relatively successful. The a6 was between a1 and b4 was between b1 and c5. There was a significant change in c1 scores, with the score gaps being 1.9641, 1.4964, and 1.2950, respectively. The trend of scores added with the imagery of the unit is plotted using the Origin software as dotted lines (Figs. 16.6, 16.7 and 16.8).

As shown in Fig. 16.6, the effect of the addition of pebble paving variables on the scoring is not significant. After adding the bamboo image, the influence of the score is more significant, and after adding a unit mural, the degree of influence on the score is less than adding a unit bamboo. In Fig. 16.6, the effect of the order from big to small is bamboo planting > bamboo chair > folk mural > pebble paving. In Fig. 16.7, the effect of the order from big to small is bamboo planting > old tree and famous wood protection > chuanxiong planting. In Fig. 16.8, the effect from the largest to the smallest is signage > bamboo planting. In the line charts, in addition to bamboo chair this measure, the degree of impact of landscape shaping measures are different. The order of the degree of impact is basically consistent with the assumptions put forward above.

In the course of the investigation, a number of participants stated that only when the elements are combined together will they be very beautiful (Fig. 16.6). The effect of scoring by adding a certain element alone is not very large. Therefore, the landscape shaping plan in a6 is selected; The addition of each element in the b chart has a significant effect on the score; a unit of bamboo added, we can see this unit of bamboo in b3 and in a5, and is only photographed from two angles. As a result, the b4 plan was selected in group b photographs. Most of the participants in c map indicated that adding a few units of bamboo had little or no effect on the scoring. Some of the participants even selected lower scores with the increase of bamboo units. Therefore, the c3 scheme was selected in the c group of photos. The determination of the three

Table 16.4 Picture classification and control variable embodiment of measures

S/N	Inside the Courtyard (a)	Inside the Courtyard (b)	Outside the courtyard (c)
1	a1: Original picture	b1: Original picture	c1: Original picture
2	a2: a1 + 3 Bamboo benches	b2: b1 + Pebble paving	c2: c1 + Signature sign
3	a3: Bamboo bench + Pebble paving	b3: b2 + 1 Block of bamboo	c3: c2 + 1 Block of bamboo
4	a4: a3 + 1 Block of bamboo	b4: b3 + Chuanqiong planting	c4: c3 + 1 Block of bamboo
5	a5: a4 + 1 Block of bamboo		c5: c4 + 3 Blocks of bamboo
6	a6: a5 + Traditional folk mural		

Table 16.5 a1–a6 scoring average, case number, standard deviation and variance statistics

Picture	a1	a2	a3	a4	a5	a6
Scoring average	4.7194	5.2518	5.2770	5.9101	6.4158	6.6835
Number of cases	139	139	139	139	139	139
Standard deviation	1.83772	1.74702	1.60747	1.55369	1.48423	1.61419
Variance	3.377	3.052	2.584	2.414	2.203	2.606

Table 16.6 b1–b4 scoring average, case number, standard deviation and variance statistics

Picture	b1	b2	b3	b4
Scoring average	4.4173	4.8849	5.5432	5.9137
Number of cases	139	139	139	139
Standard deviation	1.66876	1.67388	1.60669	1.69924
Variance	2.785	2.802	2.581	2.887

Table 16.7 c1–c5 scoring average, case number, standard deviation and variance statistics

Picture	c1	c2	c3	c4	c5
Scoring average	5.2590	5.7770	6.2158	6.3201	6.5540
Number of cases	139	139	139	139	139
Standard deviation	1.49438	1.40802	1.43463	1.53227	1.54836
Variance	2.233	1.983	2.058	2.348	2.397

Fig. 16.6 The change trend of the value of (a) with the increase of the factor. *Image source* the author

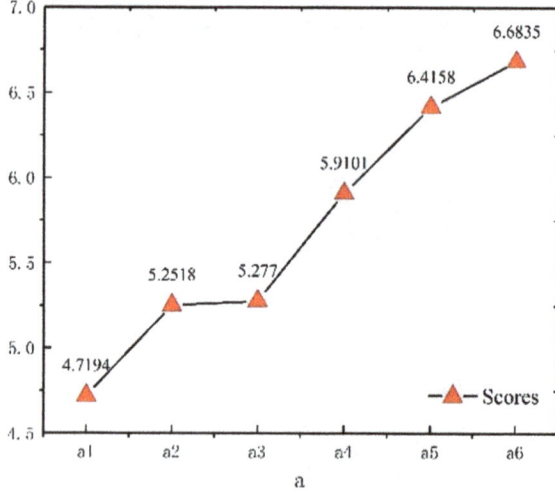

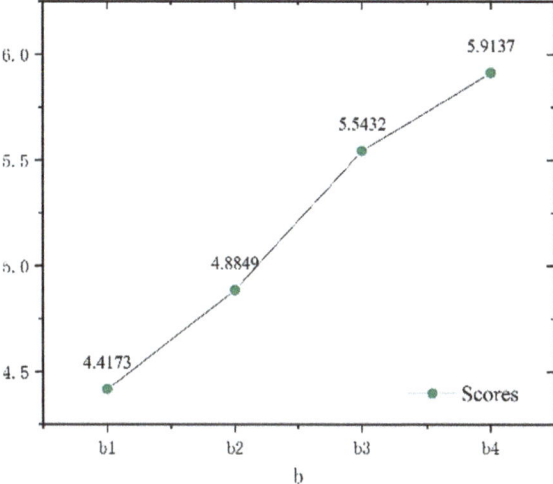

Fig. 16.7 The change trend of the value of (b) with the increase of the factor. *Image source* the author

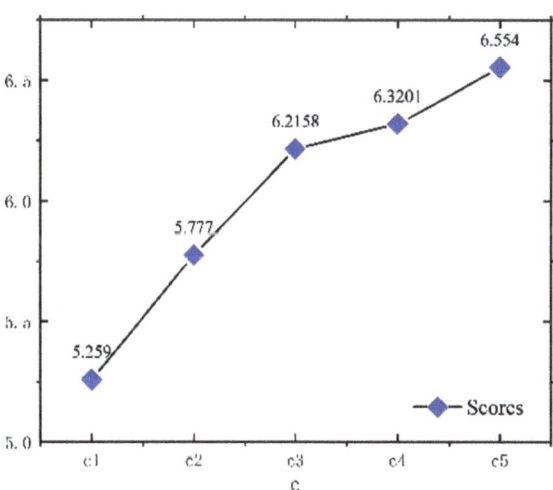

Fig. 16.8 The change trend of the value of (c) with the increase of the factor. *Image source* the author

schemes simultaneously ensures that the evaluation scores are close to or not less than six and then the lowest cost can be obtained with the aesthetic effect.

For the interior and exterior of the courtyard, the traditional building area is about 150m², so the courtyard area is about 250 m². Four bamboo chairs are added in the courtyard. The length of the bamboo identification plate is about 2 m, and the mural area is about 2 m². The area of Chuanxiong planting is about 30 m². Therefore, the cost (C1) is calculated as:

$$C(a) = 2 * C1 + 4 * C2 + 2 * C3 + 250 * C5 + 2 * C8$$
$$= 2 * 397 + 4 * 76 + 2 * 850 + 250 * 80 + 2 * 800$$

$$= 24,398 \tag{16.1}$$

$$C(b) = 2 * C5 + 30 * C7/667 = 2 * 582 + 30 * 875/667 = 1,203 \tag{16.2}$$

$$C(c) = C1 + 2 * C3 = 397 + 2 * 850 = 2,097. \tag{16.3}$$

The total cost is:

$$C1 = C(a) + C(b) + C(c) = 24398 + 1203 + 2097 = 27,698. \tag{16.4}$$

This is one of the best solutions. However, there are many possibilities. If planners want to continue to save costs on this basis, then the option of pebble paving may not be added, so the cost (C2) at this time is:

$$C2 = C1 - 250 * 80 = 7,698. \tag{16.5}$$

This is a relatively low price for the shaping plan of a village's rural cultural landscape heritage.

There is another situation that needs to be discussed. If a traditional western Sichuan dwelling has already been destroyed, we need to repair the whole house. This is the basic shaping of the rural cultural landscape heritage. Therefore, the cost (C3) at this time is:

$$C3 = C1 + 650 * 150 = 125,198. \tag{16.6}$$

If no pebble pavement is installed, the cost (C4) is:

$$C4 = C3 - 250 * 80 = 105,198. \tag{16.7}$$

For a traditional building restoration, the minimum price is ¥105,198. The actual situation is determined by the degree of historical protection of the building, so the solution to the above variables is not taken into account.

16.4 Discussion

16.4.1 The Shaping of the Linpan's Settlement Landscape Formed by Gravity Irrigation in Western Sichuan

Based on the analysis and calculation, this paper summarized the characteristics of Shiyang town's rural cultural landscape. These characteristics are dynamic, cyclical, heterogeneous, functional and cultural relics. Based on these characteristics, the appropriate shaping measures were selected to successfully shape the Linpan's settlement cultural landscape heritage. After the shaping, the typical courtyards' cultural landscape heritage also has high evaluation results and local rural characteristics. The result of shaping rural cultural landscape heritage solved many problems, such as the lack of rural cultural identity. Also, the rural style tends to be homogenized, and the traditional rural settlement cultural heritage vanishes. This also solved the problem of functional positioning of the rural cultural landscape heritage in the new era. Through the shaping results, the traditional self-flowing irrigation culture landscape heritage in western Sichuan was inherited.

Compared with the previous research results of historical and cultural village landscape protection (Peng, 2017; Guo, 2013; Jin, 2011; Li, 2011; Zhang, 2017; Fang, 2012; Fan, 2009; Xue and Zhu, 2013; Jing-Qi, 2016), this paper has put forward some suggestions and measures to protect the cultural landscape heritage, these suggestions and measures have achieved the goal of protecting the cultural landscape heritage. The differences between this paper and previous literature are reflected in the further study of the protection and shaping based on the meticulous summarization and analysis of the predecessors' work (Sun et al., 2008). The research motivation is to emphasize the necessity of combining practical projects with academic research and the significance of introducing the actual variable of cost in realizing the research results.

16.4.2 Cost Estimate of the Linpan's Settlement Landscape Formed by Gravity Irrigation in Western Sichuan

Based on the actual problem, the optimal solution model and the cost estimation of the self-flow irrigation settlement landscape of Linpan in the western Sichuan is derived and various results are obtained, making the research results more adaptive and feasible. The optimal solution obtained in this paper is to find the most suitable balance between the economic situation, the use of local materials and the high ornamental value. This paper attempts to refine and generalize the research elements of systematic and comprehensive research objects and find a suitable "bridge" between the difficult-to-quantify objects and rational analysis. In the preview articles (Guo, 2013; Jin, 2011; Peng, 2017), the status quo was often investigated and summarized. However, the perceptual suggestions put forward in the existing literature often hamper the fact that the actual project factors cannot be realized which leads to the disconnection between the research results and practical construction. This paper provides a more feasible approach to the shaping of rural cultural landscape heritage. At the same time, the research ideas of this paper also avoid the shortcomings of the relevant literature on saving costs for the observational effect and cultural heritage protection (Cortignani et al., 2018; Feng, 2014; Hideshima & Kobayashi, 1997; Hu et al., 2017). Through the process of shaping the local rural cultural landscape heritage of Linpan's settlement with the combination of different landscape images, this paper found that only through the combination of various improvements can the best effect be achieved, and that the influence of different shaping methods on the value of cultural images is significantly different. The relationship between these different measures provides an empirical basis for the selection of the focus of the future shaping program.

16.5 Conclusion

Based on modeling the rural cultural landscape heritage of typical courtyards by the optimal solution concept model, we concluded the following:

(1) The key images extracted from the rural cultural landscape image of the study area are: bamboo imagery, imagery of folk houses in western Sichuan, pebble imagery, Chuanxiong imagery, Mazu culture, and Madame Huarui;
(2) The minimum estimated cost for satisfying the good viewing effect is between ¥7698 to ¥125,198.

The optimal solution conceptual model proposed in this paper has made active and effective attempts and explorations on the combination of perceptual analysis and rational analysis, which helps academic research to better solve practical problems. In the future, the following will beare studied in depth: increasing the depth of data mining and mathematical analysis, applying more effective analysis methods, improving the optimal solution model, making the conceptual model more applicable;, and summarizing the rural cultural landscape heritage. In the future, the relationship between variables will be studied more comprehensively and systematically.

Acknowledgements This work was supported by the grants of Guang'an Vanguard District Beautiful Village Construction Technology integration and demonstration project (2016FP0127).

References

Chen, Y. (2012). *Character Assessment and Planning of Rural Landscape*. Tsinghua University.
Cortignani, R., Gobattoni, F., Pelorosso, R., Ripa, M. N., & Rosen, M. A. (2018). Green payment and perceived rural landscape quality: a cost-benefit analysis in central italy. Sustainability.
Fan, Y. (2009). *Research on Environmental Landscape Plan and Design with Safety Action of Linpan in Chengdu Plain*. Sichuan Agricultural University.
Fang, Z. (2012). *Basic Study on the Linpan Culture at Western Sichuan Plain*. Chongqing University.
Feng, N. (2014) From the perspective of economic costanalysis of landscape design. Tianjin University.
Guo, D. (2013). *Heritage Protection Research of Traditional Rural Cultural Landscape —Example: Ranzhuang*. Heibei Agricultural University.
Hideshima, E., & Kobayashi, K. (1997). The benefit and cost allocation of the conservation projects of rural landscape. *Papers on City Planning, 32*, 397–402.

Hu, R., Chen, D., & Yu, C. (2017). Low-cost Renovation Design Approach of Rural Landscape. *Journal of Shandong Agricultural University (natural Science Edition), 48*(5), 660–665.

Jin, G. (2011). *Study on Multi-Value System and Planning Design Control of Rural Landscape.* Shanghai Jiao Tong University.

Jing-Qi, L. I. (2016). A study on the approaches of rural revival and rural landscape protection in china. Chinese Landscape Architecture.

Li, F. (2011). Study on Protection Models Based on the Binary Attribute of Rural Cultural Landscape. *Areal Research & Development, 30*(4), 85–88.

Lu, C. (2012). *Researchers on the Struthers of plant Community of Landscape of Linpan West Sichuan.* Southwest Jiaotong University.

Meerschaert, M. (2015). Mathematical modeling. China Machine Press.

Peng, L. (2017). *A Discussion on the Classification and Evaluation System of Rural Cultural Landscape in Guangxi.* Guangxi University.

Sauer, C. (1925). The Morphology of Landscape. 2nd ed. University of California Publications in Geography, pp.19–53.

Sun, Y., Chen, T., & Wang, Y. (2008). Progress and prospects in research of the traditional rural cultural landscape. *Progress in Geography, 27*(6), 90–96.

Wang, M. (1999). Evaluation System of Sustainable Development in Dujiangyan Irrigation. *Area Journal of Hydraulic Engineering, 1999*(5), 13–18.

Willmott, W. (1989). Dujiangyan: Irrigation and Society in Sichuan, China. *The Australian Journal of Chinese Affairs, 22*, 143–153.

Xue, F., Zhu, Z., & Q. . (2013). The Research of Linpan Culture Landscape Conservation. *Chinese Landscape Architecture, 2013*(11), 25–29.

Yan, W., Xiang, W., & Yuan, L. (2017). Exploring Ecological Wisdom of Traditional Human Settlements in a World Cultural Heritage Area: A Case Study of Dujiangyan Irrigation Area, Sichuan Province. *China. Urban Planning International, 32*(4), 1–9.

Yuan, L. (2016). Regional Collaboration in Ecological Infrastructure Construction: Contemporary Inspiration from Ancient Water System Governance in Dujiangyan Irrigation Area. *City Planning Review., 40*(8), 36–43.

Zhang, D. (2017). Exploration on the Heritage of Traditional Village Cultural Landscape. 遗产与保护研究, 2(6):29–35.

Zhang, M., & Feng, H. (2016). Protection and Development of Historic Landscape of Dujiangyan Water System-Basing on the Viewpoint of Cultural Landscape. *Housing Science, 36*(10), 30–35.

Zhang, M., & Han, F. (2016). Conservation and Development of Dujiangyan Water System Cultural Landscape from the Perspective of Historic Urban Landscape. *Urban Development Studies, 23*(8), 60–60.

Zhang, M., Han, F. and Xu, D. (2016). Historical Landscape Protection and Development in Dujiangyan Irrigation System. Planners, 32(z2).

Zhang, R. (2013). *A Discussion on the Possibility of Specific Law of Rural Culture Landscape Heritage Conservation.* Zhejiang A&F University.

Chapter 17
An Argument Concerning Rural Planning in Contemporary China from the Perspectives of Law, Institutional Practice and Implementation Methods

Zhu Jiehao, Yang Fan, and Zhou Tianyang

17.1 Background and Introduction

The coordinated development of urban and rural areas is one of the strategic tasks demanded by China's new urbanization. Rural areas are becoming the focus of attention in domestic academia. Researchers in this field have adopted multiple approaches and have provided a number of different perspectives. Since the 19th National Congress of the People's Republic of China, the scope of research on rural areas has expanded. The discipline of urban–rural planning is also duty-bound for this growth.

Rural planning initiatives are abundant throughout the country. Although the rural planning process involves extensive preparation, it lacks systematic and theoretical guidance. Moreover, the groups involved in rural planning as a social activity come from complex and diverse academic backgrounds, have various motivations for participating, and their evaluation methods and standards are far from standardized. Most of the practitioners involved come from the disciplines like sociology, anthropology, geography, law, public administration and political science. They try to identify rural issues and propose solutions related to their own disciplinary objectives, and the urban–rural planning field has continued to deepen its research efforts on rural issues specific to their objectives as well. It has also been a long-standing

Notes *This research project is funded by National Natural Science Foundation of China (Approval Number: 51778436).

Z. Jiehao · Y. Fan (✉) · Z. Tianyang
Urban Planning Department, College of Architecture and Urban Planning, Tongji University, Shanghai, China
e-mail: fanyangsh@tongji.edu.cn

Z. Jiehao
e-mail: 14zhujiehao@tongji.edu.cn

practice to participate in rural construction through planning and design. However, under the dual influence of current national policies and rural realities, if the urban–rural planning disciplines intend to provide interactive platforms and governance mechanisms more effectively and actively, then they will have to learn from the public and optimize existing theoretical and methodological systems.

17.2 The Rank Relationship Between the Relevant Laws in Rural Planning

17.2.1 Evolutions of Rural Planning and Relevant Policies

Historically, during the feudal period, the type of state governance used in China was often to supplement imperial power, which "ended at the county level." Until the Qing Dynasty, the governance of rural areas by the local gentries has always played the role of "broker" between the state and the peasants (Du, 2003). The "asylum relationships" of traditional rural society during the Republic of China era was how the state controlled rural areas. After the founding of New China, the reform and opening, and the rise of township enterprises, the phenomenon of "geostrategism" emerged. With this changing trend, basic-level government and cadres played a big role in promoting rural governance (Qiu & Xu, 2004). In this new era, rural governance includes more emphasis on construction via on physical space, environmental ecology and infrastructure to achieve the purpose of social reconstruction through the management of physical space, using rural planning as a governance tool.

The "Urban–Rural Planning Law" standardized the conceptual categories of towns, townships, and villages, and it also defined the planning methods of villages which provided technical and legal assistance to rural planning in the urban–rural discipline. Rural planning, within the field of planning, also shifted from dealing with the issues of "land resources rights and interests, building reservation and demolition, clustering and scale expansion", to integrating the basic connotation of state governance under the concept of "people's livelihood" (Table 17.1).

The "National New Urbanization Plan" (2014–2020) proposed, from the prospective of top-level design, that China's rural planning should be people-centric, should pay more attention to the protection of traditional styles, and should inherit the essence of Chinese culture (Wen & Wen, 2015). Therefore, urban–rural planning is gradually becoming regarded as an important component of the state governance system and an important platform for achieving powerful state governance. The consensus among many scholars is that urban–rural planning should shift from pure physical spatial planning to a mode that focuses more on the interplay between economics, society, public policy, and ecological environment. Additionally, planning should no longer be monopolized by "elites", but instead require integration of elite planning, democratic planning and impartial planning, which would

Table 17.1 The representative government rural policies in recent years

Theme	Project/policy documents	Time	Remarks
Urban and rural co-ordination	The resolution of the party's 3rd Plenary Session of the 16th CPC Central Committee	October 2003	One of the 'five co-ordination'
New Countryside construction	"11th Five-Year" program outline	October 2005	
Ecological civilization, beautiful China and beautiful countryside	The resolution of the 18th CPC National Congress	November 2012	One of the 'five-in-one'
Rural revitalization	The resolution of the 19th CPC National Congress	November 2017	

(*Source* adapted from government documents at all levels).

enable communication and negotiation among various social interest groups. The exploration of rural planning based on this idea has only very recently taken shape.

17.2.2 The Plan-Making and Legal System of Rural Planning

The "Village and Town Planning Methodology (Trial)" promulgated in 2000 is an official technological regulation that guides the preparation of rural planning in the urban–rural planning discipline. It divides rural planning into two phases: master planning and construction planning. Similar to the methodology used in urban system planning, rural master planning involves a general deployment of site selection and corresponding construction plans in township administrative areas. The village construction plan is designed for the specific arrangement of villages, which is similar to the construction planning used in cities (towns) (Huang & Zhang, 2010).

Subsequently, the 2008 Urban and Rural Planning Law explicitly included villages in the urban–rural planning system. This new law marks China's attempt to transform the urban–rural dual planning system and to establish a guiding ideology for urban and rural co-ordination to achieve a certain degree of consistency between urban and rural areas in terms of planning and its legal basis (Chen, 2011). However, this law disregards its legal rank relation when compared with civil law, property law, and villager autonomy regulations, considering the rural areas in city administrative districts as the appendages of the urban area development. It can be inferred from "Administrative Measures for the Examination and Approval of the Urban Master Plan (Draft for Solicitation of Comments)" that rural planning under the urban–rural planning discipline is still situated as an important component in urban–rural comprehensive planning, while also acting as a supplement to it.

This state of rural planning embedded in the urban planning system reflects local government's administrative jurisdiction to use planning power to integrate the urban areas and rural areas under its jurisdiction, implementing tailored policies in rural areas for the overall development of the territory. Therefore, it is not surprising that rural planning is formulated as a detailed design that is then applied for rural settlements in system planning. While the technical aspects of rural planning can be of concern, rural planning is also a means to achieve spatial technology coordination for rural construction projects with different rural policy backgrounds, usually including special contents of village siting planning, village construction planning, Beautiful Village Projects, and the planning activities before the implementation of concrete construction (Shen, 2015).

17.2.3 The Coordination of Relevant Legal Relations and Reconstruction of Planning Power

Rural planning in China's urban–rural planning disciplines has long been based on "one law and one regulation"—the "Urban–rural Planning Law" and the "Regulations on the Planning and Construction of Villages and Towns". The latter provides an important legal foundation for construction planning and management in the vast rural areas in China. There are other related supporting laws and standards, including the "Methods for the Planning Formulation of Villages and Towns" and the "Standards for Village Planning". These laws and regulations cover all designated towns and townships except for the capital town, and they are influenced by the "Methods for Urban planning Formulation", emphasizing the planning object of "villages and market towns" (Fan & Lei, 2010).

In addition to the "Urban–rural Planning Law," "Land Management Law," "Environmental Protection Law," "Basic Farmland Protection Regulations" and "Agrarian Law" and other administrative regulations belong to the administrative department of the state, which is the basis of "administration according to law". They are all used to guide planning efforts to ensure greater consistency with the technical requirements, and to improve the management performance of government departments. Scholars have adopted two approaches to the position of rural planning. The first approach follows the path of entrusted agency relationships, which starts by describing the social and historical causes of rural social autonomy and the rule of law. This approach gives full respect to the autonomy of village-level self-governance organizations with villagers' interests in mind, emphasizing the sense of service and responsibility to villagers, and reconstructs planning power in rural areas through two-way interactions so that it can become an important mechanism of rural community governance. The second approach follows the logic of the urban–rural planning system by treating rural planning as a technical and governing tool adopted by local governments that allows them to consider the overall considerations of jurisdictions. Based on this notion, this approach aims to improve planning methods, and to look deeper into village-township system planning and village master planning (Shao, 2012; Gu, 2008; Lu, 2007; Liu, 2012; Li, 2010). However, it is necessary to solve the conflict, contradiction and coordination between planning administrative power and local autonomy when they extend into rural areas. Alternatively, there are potential and feasible new approaches that could be adopted to facilitate new types of planning, such as urban–rural planning, urban–rural comprehensive planning and land use zoning, in order to achieve a consistent and complete technical system.

Discussions concerning the possibility, feasibility and necessity of the first approach are rare. This also reflects that there is a path dependence in urban–rural planning circles when they are faced with major academic path-finding under the background of the reconstruction of authority power framework. For example, urban and rural planning management has changed from a management function of the Ministry of Housing and Urban–Rural Development to a management function of the Ministry of Natural Resources. In reality, the current system of laws and regulations in the adjustment of rural governance rights, land rights, property rights, and community affairs is still far from perfect. Such laws and regulations should have the important position to regulate the handling of public affairs in rural areas and to coordinate the division of rights and powers between rural and urban areas. There is an urgent need to strengthen legislation associated specifically with rural areas or adding relevant provisions on rural governance to existing laws (Xing, 2010). Moreover, in the era of territorial spatial planning, it is necessary for an adjustment of the relationship between administrative laws and the integration of laws and regulations that are attached to the relevant administrative authority, and the resulting integration of different types of planning for the implementation of different administrative authority.

17.3 Highlighting the Self-governance Characteristics of Rural Communities

17.3.1 Entrusting and Implementation Entities of Rural Planning

In 1982, the Constitution of China defined the village committee as a "grassroots organization." In 1987, the "Organization Law of the Villagers' Committee of the People's Republic of China (Trial)" was passed by the Standing Committee of the Sixth National People's Congress, marking the official entry of the self-governance of Chinese villagers into the institutionalization phase. Later, based on practical experience, the Standing Committee of the Ninth National People's Congress revised and passed the "Organic Law of the Villagers' Committees of the People's Republic of China" in 1998 which further improved the specific regulations concerning villagers' self-governance, and laid a foundation for the further development of the self-governance system of villagers. Moreover, it provided a solid foundation for self-governance in villages.

After years of practice, villager self-governance, based on democratic elections, democratic decision-making, democratic management, and democratic supervision, has seen considerable advances. With the Constitution as its foundation, the "People's Republic of China Villagers Committee Organization Law" as a cardinal point of reference, and the provincial "Method for Implementing the Village Committee Organization Law" and "Measures for the Election of Villagers Committees" as the core, the villager self-governance legal system, supplemented by State Council departmental rules and local government regulations on villager autonomy, has taken shape (Tang, 2006).

Empowered by the law, there are diverse participants in rural governance in China. They are divided into two groups: institutional participants and non-institutional participants. Institutional participants can also be referred to as authoritative governance subjects, and include township governments, village committees, and villager congresses. Non-institutional participants can be referred to as non-authoritative governance entities, and are mainly composed of rural non-governmental organizations, rural elites, and rural enterprises. In addition, there are grassroots villagers (Ma & Li, 2015). It can be seen, in this system design, that the planning and construction of public space should be formed by the collective resolution of villagers organized by the "two committees of the village" (Sun, 2014). In the context of the "governance of the township and the village", the township government, as the representative of authoritarianism, still affects the direction of rural development and thus villager autonomy has not been realized.

In the process of intervention in the construction and development of rural areas, the authority to regulate planning, which is part of the administrative power of the city government, has been separated from government structures at the corresponding level. This has resulted in the separation of the entrusting subject, the implementation subject and the demanding subject. Since the villagers' autonomy power is excluded from the principal part of the commission, it also leads to an absence of the villagers' rights to express their wants and needs. Therefore, in the implementation of planning efforts, who plans and who implements becomes a normal choice. Many initiatives of autonomous organizations have not been well mobilized.

17.3.2   Land Right Characteristics of Rural Community

The relationship between land ownership, contracting rights, and management rights, which is unique to China's rural areas, is completely different from the relationships that govern urban land use. As such, these land right characteristics have emerged as another important institutional factor determining and influencing rural planning.

Whether it is the Constitution, Land Management Law, or Agricultural Law, it is clearly stipulated that rural land is owned collectively by villagers. Due to the imperfections of the current rural social security system, land has an additional social security function for villagers, and its value has gradually exceeded its original function as a means of production (Huang, 2011). This utility of land has led migrant workers working in cities to treat land as inheritable property, and they would prefer to operate the land in perpetuity than lose ownership.

With the increase of rural migrant workers, the separation of people and land in rural areas is becoming more common, and the separation of contracting rights and management rights has gradually emerged as the new norm. At the same time, the small-scale farming production model, characterized by fragmented management methods, has become widely used. The development of modern agriculture, characterized by specialization, mechanization, and marketization, is incompatible with the simultaneous development of both small- and large-scale land management. The "three-rights separation" of "land ownership, contracting rights, and management rights" in rural areas provides the basic direction for the new round of reforms in China's rural land system (Chen, 2016).

With the development of rural land system reform, more villagers can work in cities while retaining their rural land. This has led to an increase in the flow of people between urban and rural areas and the growing separation of people and land in rural areas. If rural planning fails to capture the direction of the land system reform in a timely and sensitive manner, continuing to ignore the ownership characteristics of land, it would lead to another loss of discourse in proposing strategic policies.

17.3.3 The Government Promotion of Rural Construction

Current rural planning depends largely on the promotion of basic-level government, but can also be based on social stability, government management and construction directed towards the goals of economic development. Although villagers' self-governance under the constraints of the traditional clan in the countryside can present simple and creative development plans, the resources and efforts that can be mobilized are far less than are possible with the government's operation, and their results have fewer aggregation effects than those of the government. Therefore, promoting rural planning, economic development and social transformation in rural areas has become a new "public service" provided by the basic-level government, in addition to the improvement of public facilities, the resolution of contradictions, and the control of violations. When faced with a plight related to finance and capability, or common problems such as labor shortages, rural planning becomes a form of political declaration or mobilization (Wen & Wen, 2015).

As a result, rural planning from the perspective of urban–rural planning serves as an "assistant" to basic-level governments in promoting rural construction, often led by the requirements of external agents, and it usually overlooks the spatial, social, and cultural needs of the villagers. Practitioners simply extend the spatial construction methods adopted in urban areas to rural planning. Furthermore, and due to a lack of familiarity with rural issues, planning fails to take the following into full consideration: the rural social order, spatial relationships in the neighborhood, and rural socio-economic development paths. This produces various contradictions in the implementation process. When rural planning is exercised under the guidance of such external values, the internal mechanism of traditional rural space production is altered and new lifestyles are implanted (Meng, 2015).

Relevant research studies, however, tend to remain confined to a discussion of rural planning systems, formulation methods and planning modes, and often neglect the fundamental problems related to rural practices, such as the unique political context of rural life, the incompatibility between urban planning methods and rural areas, and the short-term and long-term visions of different stakeholders in rural areas. This neglect may lead to difficulties in implementing most rural planning projects.

17.4   Exploration of Spatial Language System of Rural Planning

17.4.1   Current Research and Analysis Methods

Currently, rural planning research and survey methods are identical to urban planning methods, including on-site surveys, questionnaires, and household interviews. In fact, research and analysis of rural planning needs to focus more on the characteristics of agriculture and the distinct qualities of rural areas—geographical characteristics, agriculture characteristics, social relations, and cultural features—and should adopt a vocabulary and channels (and even communication methods) that are appropriate for these features. Simply copying the investigation methods used for cities in order to understand the context of village development through detailed analyses of populations, land, and surveys of human relationship will prove to be inadequate, or even wrong.

In field research, the accuracy, completeness, and real-time nature of national land-related data (such as the translation of rural land-based information through aerial imagery), statistical scope, and the comprehensive attributes of information (use, scale, geography, space, etc.) cannot satisfy the requirements of rural planning. In the absence of solid site remapping and surveyance, it is difficult to obtain sufficient information to help planners formulate suggestions. For example, the situation and range of villager contracted land is not reflected in most cadastral information drawings. In addition, in rural areas, the behavioral logic of an individual villager and overall behavioral characteristics of communities determine the status and distribution of land use and development, and in turn affects the livelihood of the villagers (Qiao et al., 2016).

17.4.2 Problems in the Policy of Spatial Planning

In rural planning, there are several technical paths available to plan for rural communities. The first is based on urban–rural co-ordination, and involves the implementation of village development system arrangements, and the establishment of macro-to-micro-level controls and communication. For example, rural planning in the Zengcheng District of Guangzhou is carried out in accordance with the model of "county-level urban and rural development system" and "Planning of Beautiful Rural Villages". The county-level urban–rural development system focuses on the deployment of village and township systems, the layout of land use, space control, industrial development, road traffic, facilities planning, and rural landscape improvement. The beautiful village construction plan focuses on issues such as industrial construction, rural housing construction and management, rural infrastructure, environmental landscape improvement, and rural cultural construction (Jiang & Yuan, 2016). The second system relies on rural projects. For example, the Nanjing Jiangning District launched a series of projects such as "Five Golden Flowers" and "Beautiful Rural Village Model Villages". Village planning serves as a means of coordinating various projects and disassembling them into achievable engineering projects such as ecological restoration and tourism area construction (Shen, 2016). The third system is focused on village planning where the main idea is to guide the revitalization of industry, improve the environment that villagers inhabit, activate local traditional culture, and preserve historical remains. It adopts a planning expression system that divides functional areas and connects the flow of people through the construction of a "public space" system. Many believe that current rural planning is a replica of urban planning in rural areas. It is an extension and intrusion of urban planning techniques and values into rural areas without appropriate modifications.

Rural cultivation culture, neighborhood kinship relations, and rural customs are key factors that characterize "acquaintance societies", and it differentiates them from modern "stranger" societies (Wen & Wen, 2015). It is the conflicts of interest in land property rights and the complicated relationship of clan beliefs that bring these differences into the ambit of village governance and rural planning among different villages. Some scholars and experts once hoped that urban–rural planning could simply be extended to cover township and village planning, by improving urban planning and design models and adopting the methods of urban design to rural contexts. However, this runs the danger of severing rural planning completely from rural reality, and creating permanent "archives" that cannot be put into practice (Zhang, 2014).

With their professional knowledge and the technical means at their disposal, designers should have been able to grasp local realities and folk characteristics and generate rural plans that meet local needs; however, the current situation is very different. Current rural planning methods make it difficult to avoid the mispositioning of thinking modes, concept guidance, and value judgments, which generally results in impoverished local intellectual energies. Planners are not ready to communicate effectively with the villagers; the proposed plans are either not recognized and understood by the villagers, are not intended to be discussed by the villagers at all, or are just for the praise of experts and local government officials. If the goal of rural revitalization is really established through the implementation of rural planning, then the theoretical method of modern urban planning based on the western cultural system should be creatively redirected and the method should be tailored to eastern systems.

17.5   Conclusion and Prospective

The "Urban and Rural Planning Law", which is relied on in urban-rural planning, has not yet clarified its relationship with the laws and regulations such as "Constitution, property law, village group law, autonomous regulations, land law, environmental protection law" etc. including complex weighting relationship and legal rank relation. On the one hand, the village collective is not the executive power implementation tool of the basic-level government, and on the other hand, planning behavior is regarded as a type of representation for the local government power. As a result, it is difficult for rural planning to play its role in solving the problems of rural land ownership and the specific interests of villagers and village collectives. At the same time and at the level of rural planning implementation, there is also a lack of coercive power to coordinate the demands of various interest groups making it difficult to implement planning schemes.

Furthermore, whether the subject field of urban–rural planning is a simple sum of two fields of rural planning and urban planning is worthy of academic research and contention. Villages have a direct bearing on the livelihood of the village residents, and have an impact on the development of urban–rural relations (Zhou & Yu, 2014). Therefore, the following issues are needed to be answered by scholars: Should rural planning be completely dependent on the self-development vision of rural self-government communities? How can the coordination and interaction mechanisms between rural self-government and urban governance be effectively translated to rural planning and urban planning? What is the significance of rural revitalization for cities and villages? How are they achieved? How are they discussed, planned, and formed in order to be realized in shared programs of action through urban planning and rural planning? These issues will have an important impact on the actions of urban–rural planning scholars.

References

Chen, F. (2011). An approach to the rule of law in village planning under the background of urban and rural coordination. *Journal of Suzhou University (social Science Edition), 02*, 115–119.
Chen, C. (2016). Three-power distribution" in rural land: functional roles, division of powers, and institutional construction. *China Politics & Resources, Environment, 26*(4), 135–141.

Du, Z. (2003). Culture, power, and country: North China rural areas 1900–1942: 1900–1942 NIAN DE HUA BEI NONG CUN. [2] Edition. Jiangsu People's Publishing House.

Fan, L., & Lei, C. (2010). On the legal implementation strategy of rural planning in China: Based on the discussion of urban and rural planning law. *The Planner, 26*(1), 5–9.

Gu, C., & Wu, L. (2008) A summary of the main achievements of china's urbanization research. *Urban Issues* (12), 2–12.

Huang, K. (2011). Influence of rural land system on the urbanization of new generation migrant workers and institutional innovation. *Research of Agricultural Modernization, 32*(2), 196–199.

Huang, X. &Zhang, X. (2010). A preliminary study on the construction of new rural planning system in the background of the integrated development of urban and rural areas: A case study of Wuhan City. *Planner* 26(07), 76–79.

Jiang, W., & Yuan, N. (2016). A study on the compilation of county rural construction planning pilots: A case study of Zengcheng District of Guangzhou. *Guangdong Province. Small Town Construction, 6*, 33–39.

Li, W. (2010). Research on the construction of content system of village planning. *Suzhou University of Science and Technology*, 2010.

Liu, S. (2012). Discussion on the planning of village planning in Towns with Dense Areas – Taking Guangzhou as an Example. // *2012 China Urban Planning Annual Conference* 2012.

Lu, X., & Sun, M. (2007). Urban and rural planning system adaptation based on innovation and development of rural areas in south of Jiangsu Province. *City Planning, 31*(7), 73–76.

Ma, C., Li, X., et al. (2015). Development logic and realization path of rural multi-subject cooperative governance. *Journal of Shanxi Agricultural University (social Science Edition), 14*(7), 674–678.

Meng, Y., Dai, S., Wen, X. (2015). The problems and countermeasures faced by the current rural planning practice in China. *Planner, 2*,143–147.

Qiao, J., Hong, L., & Wang, Y. (2016). Levels and logic of rural planning in the context of ecology and human context: Based on the investigation and practice of mountainous areas in Western Hubei Province. *Urban Development Research, 23*(6), 88–97.

Qiu, H., & X, J. (2004). Local government behavior in technological innovation of industrial clusters. *Management World*, (10), 36–46.

Shao, L., & Zhou, D. (2012). Jinan urban and rural overall planning system for planning authority. *The Planner, 28*(4), 46–51.

Shen, M. (2015). Rural governance driven by rural projects and planning: An empirical study based on Nanjing Jiangning. *City Planning, 39*(10), 83–90.

Sun, Z. (2014). Study on the governance structure of rural China in contemporary China: Current situation and analysis. China Social Sciences Net.

Tang, M. (2006). Two basic issues concerning the improvement of the legal system of villagers' autonomy. *Journal of Legal Research, 2*, 3–8.

Wen, J., & Wen, W. (2015). Research on the problem of rural governance and planning in China. *Modern Urban Research, 4*, 16–26.

Xing, H. (2010). Characteristics of rural land use planning laws insome countries and regions and their use for reference. *International Urban Planning, 25*(2), 26–30.

Zhang, S. (2014). Rural planning: Characteristics and difficulties. *City Planning*,.318(2), 17–21

Zhou, L., & Yu, C. (2014). The international experience of rural planning and construction and the professional thinking in Jiangsu Practice. *International Urban Planning, 29*(6), 1–7.

Lightning Source UK Ltd.
Milton Keynes UK
UKHW020006231122
412632UK00002B/12

9 783030 838584